AF865987

Emerging Trends in Wireless Sensor Networks

Edited by Venkata Krishna Parimala

Published in London, United Kingdom

Emerging Trends in Wireless Sensor Networks
http://dx.doi.org/10.5772/intechopen.95653
Edited by Venkata Krishna Parimala

Contributors
Ki-Il Kim, Beom-Su Kim, Babar Shah, Tareq Alhmiedat, Nadir Hakem, Nour Zaarour, Nahi Kandil, A. Pravin Renold, B. Venkatalakshmi, Seifeddine Messaoud, Soulef Bouaafia, Abbas Bradai, Mohamed Atri, Mohamed Ali Hajjaji, Abdelatif Mtibaa

First published in London, United Kingdom, 2022 by IntechOpen
IntechOpen is the global imprint of INTECHOPEN LIMITED, registered in England and Wales, registration number: 11086078, 5 Princes Gate Court, London, SW7 2QJ, United Kingdom

British Library Cataloguing-in-Publication Data
A catalogue record for this book is available from the British Library

Additional hard and PDF copies can be obtained from orders@intechopen.com

Emerging Trends in Wireless Sensor Networks
Edited by Venkata Krishna Parimala
p. cm.
Print ISBN 978-1-80355-219-4
Online ISBN 978-1-80355-220-0
eBook (PDF) ISBN 978-1-80355-221-7

Meet the editor

Dr. Venkata Krishna Parimala is currently a Professor of Computer Science and a director for university rankings at Sri Padmavati Mahila University, India. He received his BTech in Electronics and Communication Engineering from Sri Venkateswara University, India, an MTech in Computer Science and Engineering from Regional Engineering College, India, and a Ph.D. in Computer Science and Engineering from VIT University, India. Dr. Krishna has several years of experience in academia, research, teaching, consultancy, academic administration, and project management. His current research interests include mobile and wireless systems, data science, quality of service (QoS), and cloud computing. He is the recipient of several academic and research awards such as the 2009–2010 Cognizant Best Faculty Award and the 2021 Best Researcher Award from Sri Padmavati Mahila University. Recently, Prof. Krishna was listed among the top 2% of scientists in the world by Stanford University. He has authored more than 200 research papers in various national and international journals and conferences. He has delivered several keynote addresses and chaired sessions in reputed conferences. He is a senior member of several professional societies such as the Institute of Electrical and Electronics Engineers (IEEE), Association for Computing Machinery (ACM), Computer Society of India (CSI), Institution of Engineers (India), and others.

Contents

Preface

The importance of wireless sensor networks (WSNs) has been increasing every day since the introduction of the Internet of Things (IoT). The usage and applications of WSNs are expanding in the fields of automobiles, defense technologies, medical systems, agriculture, weather monitoring, and more. A WSN consists of four major components: sensors, transmitters, an access point (AP) or receiver, and an evaluation node. Sensors capture data and transmitters forward the data received from the AP. Upon successful retraction of data at the AP, the evaluation component generates the data for the end user.

WSNs play a major role in the growth and success of wireless and IoT systems and applications. However, the deployment of WSNs does have challenges and limitations. The performance of a WSN greatly depends on the energy requirements of the applications. Some of these IoT applications include Smart Traffic Monitoring Systems (STMS), Water Quality Monitoring (WQM), Air Quality Index (AQI), and so on. For example, a simple STMS uses sensors that capture images of traffic and send them to the WSN for further processing, after which the traffic signal is modified.

This book addresses various aspects of WSNs, including novel strategies based on wireless systems and IoT, such as machine learning, optimization, control, statistics, and social computing, among others.

Section 1 includes Chapter 1 "Localization Context-Aware Models for Wireless Sensor Network" by Hakem et al. This chapter provides details about context models for WSNs, which are gaining traction as a critical technology for enabling IoT and smart objects. Small devices with low energy consumption and limited computing resources are widely used in a range of applications and fields. Nodes are distributed at random with no prior knowledge of their location. However, location context is a critical feature for providing a context-aware framework to information gathered from sensors in a variety of services such as intrusion detection, surveillance, geographic routing/forwarding, and coverage area management. This chapter examines several existing cutting-edge technologies.

Section 2 begins with Chapter 2, "IoT-Based Decision Support System for Health Monitoring of Induction Motors", by Renold and Venkatalakshmi, which proposes an IoT-based solution for continuously monitoring and recording induction motor vibration. An electrical motor is a common device used for a variety of electrical applications. Motors that are both reliable and long-lasting are in high demand. Motors are susceptible to faults such as rotor bar breaking faults, short turn faults, bearing outrace faults, and so on. These motors' unexpected faults or failures reduce workplace productivity. The shape of most electrical motors with rolling bearings becomes disproportional due to continuous operation. This causes the elastic limit of the motor to be exceeded, as well as fractures, vibrations, and a temperature increase.

Chapter 3, "Network Slicing for Industrial IoT and Industrial Wireless Sensor Network: Deep Federated Learning Approach and Its Implementation Challenges", by Seifeddine Messaoud et al., examines issues related to network slicing. Network slicing is being recognized as a beacon technology that enables multi-service Industrial Internet of Things (IIoT) networks. This chapter discusses the new federated learning paradigm before proposing a Deep Federated RL (DFRL) scheme to provide federated network resource management for future IIoT networks.

The final section of the book discusses WSN management. Chapter 4, "Resource Allocation in Wireless Body Area Networks: A Smart City Perspective", by Kim et al., discusses health care as an essential service in smart cities. Personal health monitoring systems, infrastructure for collecting and delivering individual data, and a system for diagnosing symptoms are required to deploy healthcare systems in such cities. Wireless body area networks (WBANs) have recently received a lot of attention from the research community for the first requirement. As such, this chapter presents a thorough discussion of emerging research trends in WSNs, as well as a discussion of WBANs in terms of resource allocation. Finally, Chapter 5, "Energy Management in Wireless Sensor Network", by Alhmiedat et al., discusses the importance of energy management and presents possible energy management strategies to overcome energy depletion in sensor nodes. WSNs in large areas are used to monitor various physical conditions of the environment and transmit the collected sensed data to a base station (central node). Some examples include gas monitoring, intrusion detection, object tracking, and so on. Sensor nodes are typically unattended and battery-powered with no external power source. As a result, WSNs face the challenge of limited onboard energy sources, where packet transmission and sensing functions consume the most power.

I would like to thank Ms. Ana Javor and other members of the editorial team at IntechOpen for their kind cooperation and help. I also extend sincere thanks to the contributing authors and reviewers for their interest and support.

Venkata Krishna Parimala
Computer Science Department,
Sri Padmavati Mahila University,
Tirupati, India

Section 1

Wireless Sensor Network Models

Chapter 1

Localization Context-Aware Models for Wireless Sensor Network

Nour Zaarour, Nadir Hakem and Nahi Kandil

Abstract

Wireless sensor networks (WSNs) are emerging as the key technology to support the Internet of Things (IoT) and smart objects. Small devices with low energy consumption and limited computing resources have wide use in many applications and different fields. Nodes are deployed randomly without a priori knowledge of their location. However, location context is a fundamental feature necessary to provide a context-aware framework to information gathered from sensors in many services such as intrusion detection, surveillance, geographic routing/forwarding, and coverage area management. Nevertheless, only a little number of nodes called anchors are equipped with localization components, such as Global Positioning System (GPS) chips. Worse still, when sensors are deployed in an indoor environment, GPS serves no purpose. This chapter surveys a variety of state-of-the-art existing localization techniques and compares their characteristics by detailing their applications, strengths, and challenges. The specificities and enhancements of the most popular and effective techniques are as well reported. Besides, current research directions in localization are discussed.

Keywords: WSN, localization, GPS, range-free, range-based, anchors, mobile nodes, 3D localization

1. Introduction

The popularity of wireless sensor networks (WSNs) is taking advantages of advances in wireless communication and digital electronics [1]. WSN have become an important and interesting subject of research. It is composed of small devices equipped with a microcontroller, a low-power radio, and a number of sensors to observe the environment. The Internet of Things (IoT) is defined as "Simply, the Internet of Things is made up of devices – from simple sensors to smartphones and wearables connected together," Matthew Evans, the IoT program head at techUk, says [2]. Hence, localization-based services are the most important issues related to the IoT.

Smart environments constitute an evolutionary development step in many applications and fields, such as tracking, environmental monitoring, disaster management, climate control, health care, human monitoring, and underground mining. Node localization is essential to provide a physical context to sensor readings for services such as intrusion detection, surveillance, geographic routing, and coverage area management [3].

Localization, also known as positioning problem, is a one-time detection technique, where the position or the location of the unknown sensor is estimated. The closeness of the estimated location to the real value presents the accuracy of the technique, and the consistency of the estimated location presents the precision of the technique. However, a sensor location can be global or relative. A global position is provided by a global reference such as the Global Positioning System (GPS) or the Universal Transverse Mercator (UTM) coordinate system. On the other hand, relative position is based on an arbitrary coordinate system, for instance, a sensor's location is obtained as distances to other sensors. Tracking is an on-time method where the trajectory of an unknown sensor is estimated in real-time applications [4], also known as connectivity, which indicates whether two sensors can communicate between them through one hop, that is, a packet transmitted by one sensor can be received by the other sensor. Reference nodes, also known as anchor nodes, are aware of their positions in the network, they are used by unknown nodes in the localization process. Techniques based on anchors are anchor-based, the estimated positions are global metrics, otherwise, the technique is anchor-free. Localization techniques based on measurements such as distances or angles between sensors are called range-based techniques, as opposed to the range-free techniques [3]. This chapter will discuss all these different techniques as well as their concepts, drawbacks, and advantages, as well as presenting briefly localization applications.

2. Global positioning system localization

The GPS is one of the well-known and used among global navigation satellite systems (GNSS). Owned and operated by the U.S. government, GPS provides global coverage. Using this system and with respect to a reference in time and space, users estimate accurately and in real time their three-dimensional (3D) position, velocity, and time [5]. It consists of at least 24 satellites, arranged in six orbital planes with four satellites per plane, orbiting the earth at altitudes of approximately 11,000 miles. An unlimited number of users can be positioned by GPS, by using the concept of a one-way time of arrival (TOA) ranging. The distribution of satellites ensures that at least eight satellites can be seen simultaneously from almost anywhere on the planet. Each satellite broadcasts waves containing information on its identity, its location, and the date and time the signal has been sent. These waves propagate at the known speed of light. The GPS receiver receives the information transmitted by the satellites and determines the time difference between the code generation time and the reception time. Then, the distance separating the satellite to the receiver can be calculated by a simple relation between speed and time (distance $=$ speed $\times$ time). Using this distance, the receiver is said to be located on a sphere centered on the satellite with a radius equal to the computed distance. This process is repeated with two more satellites, then the position of the receiver is estimated as the intersection of three spheres [3].

However, in WSNs fully GPS-based solution is impractical, since not each sensor can have its own GPS receiver. This is due to many constraints such as cost, high-power consumption, and the need for line-of-sight between GPS and satellites. Also,

DOI: http://dx.doi.org/10.5772/intechopen.103893

GPS performance will deteriorate considerably when deployed in hostile or very severe environments [5]. In addition, if indoor scenarios are considered, the GPS signal will become even worse and therefore unreliable for location. To locate nodes in networks of mobile sensors in larger and/or mobile networks, various techniques and localization algorithms have been proposed.

3. Localization context-aware applications

The increasingly reduced size of sensors, their low cost, the wide range of types of available sensors, as well as the wireless communication medium used allow WSN to quickly invade several fields of application. The diversity of applications is remarkable, among the fields where they can offer the best contributions, we can cite the following fields: military, environmental, health, security, underground mining, etc.

3.1 Military application

WSN can rapidly be deployed and used for military applications such as battlefield surveillance, combat monitoring, and intruder detection [6]. The main advantage of using WSN is their capacity to be spontaneously positioned since the terrain of the battlefield is variable [6]. In addition, enemy location can be expected to use WSN in combat monitoring, which is the most critical information to detect intruders [3].

3.2 Health application

Using WSN in the healthcare domain allows providing real-time positioning of patients (patients with Parkinson's disease, epilepsy, patients with heart disease, and the elderly) in a hospital or their homes by using wearable hardware for example [7].

3.3 Environmental application

Air monitoring, water monitoring, and emergency alerts are subcategories of environmental applications. An important aspect is proactive monitoring of common disastrous causes in real time to lower or prevent damage [6]. For example, the integration of sensors in the walls promotes the detection of alterations in the structure of a building following an earthquake or aging, and the monitoring of movements in order to constitute a system of detection of distributed intrusions.

3.4 Underground mining application

The underground mining environment is one of the most dangerous working environments. Many accidents occur in mines causing death and loss of several people, and this is due to a lack of surveillance and detection of danger. WSNs make working conditions easier and safer, and also facilitate rescue operations. In fact, sensors are deployed to locate people in normal or abnormal situations such as accidents. Moreover, sensors can be used to locate holes that cause collapses [3]. However, the mining environment is hostile for radio communications, which cause several challenges during the deployment of the WSNs in underground mines; also, the signals reach the destination after having undergone several physical phenomena, such as reflection, refraction, and dispersion. Besides, due to the high percentage of relative humidity, signal absorption and attenuation are extremely high. Therefore, the deployment of WSNs in an underground mine must take into account a compromise between the contradictory requirements [3].

4. System model

Consider a WSN with N nodes randomly deployed. Each node $i, i \in \{1, 2, \dots, N\}$ is characterized by its physical position p_i given by:

$$p_i = \begin{cases} [x_i](1\text{D}) \\ [x_i, y_i]^T(2\text{D}) \quad i = \{1, 2, \dots, N\} \\ [x_i, y_i, z_i]^T(3\text{D}) \end{cases} \tag{1}$$

The purpose of the localization is to compute the unknown vector p_i. A WSN can be modeled as a graph G:

$$G = (V, E) \tag{2}$$

where $V = \{1, 2, \dots, N\}$ is a set of vertexes that contains an element for each node, while the set E contains an edge $\{i,j\}$, where i and j are neighbors, that is, they can exchange radio messages within one hop [8]. Hence, the localization problem is analog to the problem of embedding a graph in a Euclidean space, and that by finding a mapping function f such that:

$$f : V \to \mathbb{R}^d \tag{3}$$

This function uses constraints derived from the edge to assign each vertex to a position in $\mathbb{R}^d$, with d the dimensionality of the embedding space [8].

4.1 Embedding with known edge lengths

Measurements m_{ij} are available and are estimates of the distance between two nodes i and j, when some of the inter-node distances are known. Thus, the embedding problem searches a mapping f compatible with the obtained data:

$$\| f(i) - f(j) \| = m_{ij}, \forall \{i,j\} \in E \tag{4}$$

where $\|.\|$ denotes the Euclidean norm. This approach is used in case of range-based techniques.

4.2 Embedding using connectivity information

A different approach is used by range-free schemes that only rely on connectivity information. The model of a network with connectivity constraints can be represented as an idealized wireless network, where two nodes are neighbors if and only if their distance is less than the communication range R of nodes.

5. Range-based localization

Range-based schemes are derived from distance and angle estimation techniques. They use the distance/angle between sensors to estimate the location. The technique accuracy depends on the quality of signal measurements. However, range-free techniques are based on the connectivity to estimate the position of a sensor relative to other sensor nodes. Range-free techniques are cost-effective solutions; however, the accuracy is lower than the accuracy of range-based techniques.

5.1 Ranging techniques

Information on distances or angles can be obtained using many techniques, such as the received signal strength (RSS) [9, 10], the ToA [11, 12], the time difference of arrival (TDoA) [13, 14], and the angle of arrival (AoA) [15, 16]. Range-based techniques have a high accuracy range compared to range-free techniques. However, they require additional hardware making them expensive for large systems. **Table 1** summarizes some range-based algorithms as well as their pros and cons.

5.1.1 Time of arrival

Also called time of flight, ToA is a timing-based technique that depends on accurate measurements of transmitting and receiving time of signals between two nodes. Based on the known speed of the signal (acoustic signal travels at a velocity of 343 m/s and radio signal at a velocity of 300 km/s) and on propagation time obtained from these measurements, the distance separating these nodes is calculated [1]. ToA requires highly accurate synchronization of the clocks of the sender and receiver at the microsecond level. All signals transmitted must incorporate a timestamp to accurately estimate the distance traveled. There exist two types of ToA techniques [3]. The one-way ToA method measures the difference between the sending time and the signal arrival time. It requires highly accurate synchronization of the clocks of the sender and receiver (**Figure 1a**). The distance between two nodes i and j can be calculated as:

$$d_{ij} = \tau \times v \tag{5}$$

Method	Pros	Cons
AHLoS	• Acceptable accuracy in small-scale networks.	• Poor accuracy in large-scale networks
TPS	• Independent TDoA measurements required.	• Powerful anchors required (may not be valid in WSN).
MAL	• Simple computation. • No additional distance measurements required.	• Large latency in large networks.

Table 1.
Range-based algorithms.

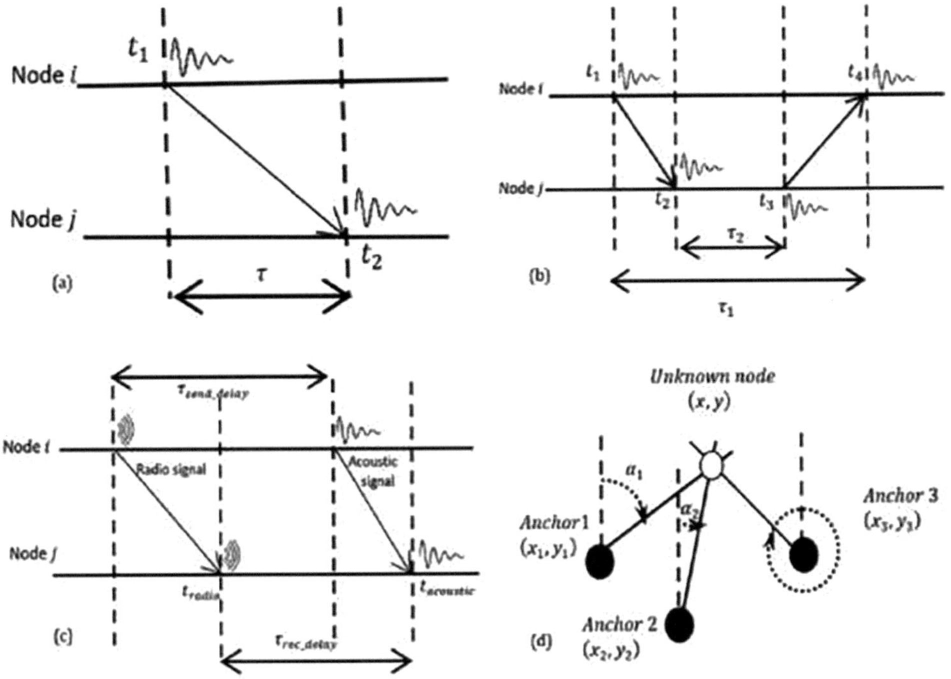

Figure 1.
Ranging techniques: (a) one-way ToA, (b) two-ways ToA, (c) TDoA, and (d) AoA.

where $\tau = t_2 - t_1$, t_1 and t_2 are the sending and receiving times of the signal respectively, and v is the signal velocity.

The two-way ToA method is preferred, where the round-trip time of a signal is measured at the sender device (**Figure 1b**). The distance is calculated as:

$$d_{ij} = \frac{\tau_1 - \tau_2}{2} \times v \tag{6}$$

With $\tau_1 = t_4 - t_1$ and $\tau_2 = t_3 - t_2$, t_3 and t_4 are the sending and receive times of the response signal, respectively. ToA techniques require very accurate hardware to measure the actual received time of the signal by the nodes. Since the propagation velocity of RF signals is high, any small error in the time measurement results in a large distance estimate error. Thus, ToA techniques are not practical for traditional WSNs [1, 3].

5.1.2 Time difference of arrival

TDoA technique is also a timing-based technique that uses two separate signals traveling with different velocities (like radio/ultrasound or radio/acoustic). The difference between their receive times can be used to estimate the distance between nodes. This approach defines a hyperbolic area where a target is possibly located with two paired sensors as foci [17]. Each node is equipped with a microphone and a speaker. The transmitter (anchor node) sends a radio message and waits some fixed interval of time t_{delay}.

Then, it produces a fixed pattern of "chirps" on its speaker. When a node receives the radio signal, it notes the current time (t_{radio}) and turn on its microphone. When this latter detects the chirp, the node notes again the current time ($t_{acoustic}$) (**Figure 1c**). Knowing t_{radio}, t_{delay}, and $t_{acoustic}$ and given that radio waves (with velocity v_{radio}) travel faster than sound in air (with velocity $v_{acoustic}$), distance d_{ij} between two nodes can be calculated as in [18] by Eq. (7).

$$d_{ij} = (v_{radio} - v_{acoustic})\left(t_{acoustique} - t_{radio} - t_{delay}\right) \tag{7}$$

TDoA based approaches do not require synchronization between the clocks of the sender and the receiver. However, it needs additional hardware (i.e., microphone, speaker, etc.) [1, 3].

5.1.3 Angle of arrival

The direction of the received signal can also be used for localization. The AoA is defined as the angle between the propagation direction and some reference direction known as orientation (**Figure 1d**) [19]. Data is collected using radio or microphone arrays. The AoA of the signal (α_i) is calculated by studying the time difference or the phase between the signal's arrivals at different microphones. The relationship between coordinates of an unlocalized sensor (x, y), anchors' coordinates $(x_i, y_i), (i = 1, 2, \ldots, n)$ and angles of arrival α_i can be expressed by Eq. (8).

$$d_{ij} = (v_{radio} - v_{acoustic})(t_{acoustic} - t_{radio} - t_{delai}). \tag{8}$$

Knowing the angles of arrival from two or more anchors, sensor's location can be estimated using a standard least-squares approach.

$$AX = b, X = \begin{bmatrix} x \\ y \end{bmatrix}. \tag{9}$$

$$A = \begin{pmatrix} 1 & -\tan\alpha_1 \\ 1 & -\tan\alpha_2 \\ \vdots & \vdots \\ 1 & -\tan\alpha_n \end{pmatrix} \quad b = \begin{pmatrix} x_1 - y_1\tan\alpha_1 \\ x_2 - y_2\tan\alpha_2 \\ \vdots \\ x_n - y_n\tan\alpha_n \end{pmatrix} \tag{10}$$

Hence, estimated location, $\hat{X}$ is calculated by:

$$\hat{X} = \left(A^T A\right)^{-1} A^T b \tag{11}$$

Depending on the measurements, AoA techniques, directionally based techniques, provide high localization accuracy. Nevertheless, higher complexity antenna arrays are essential for measurement, increasing thus the cost of WSN. Moreover, to offer spatial diversity and to measure accurately the AoA, a space is required; however, it may not be possible in WSN, considering the size of sensor nodes. Besides, these techniques suffer from multipath and scattering as well as NLoS conditions.

5.1.4 Received signal strength indicator

The most used range-based technique is based on RSS measurements. Each node is equipped with a radio. Based on theoretical or empirical models, the distance between two nodes is estimated based on the power of the received signal [1]. Wireless network card drivers export received signal strength indicator (RSSI) values, but the relationship between RSSI values and the signal's power levels differ from brand to brand [3]. In theory, the energy of a radio signal decays with the square of the distance from the signal's source [20]. The signal strength measured by a receiver at a given distance d can be calculated as in Eq. (12).

$$\begin{aligned} P_r(d) &= P_t + G_t + G_r - \overline{PL}(d_0) - 10\,n \log\left({}^{d}\!/\!_{d_0}\right) + X_\sigma \\ &= P_0 - 10\,n\,\log\,(d/d_0) + X_\sigma \end{aligned} \tag{12}$$

$P_r(d)$ represents the RSS, P_t the transmitted power, and G_t and G_r the gain of the transmitter and receiver antenna, respectively. The constant term P_0 represents in fact the RSS measured at a distance d_0 (reference distance), n is the path loss exponent, and X_σ is the uncertainty factor due to multipath and shadowing [1]. Whereas, the accuracy of this ranging technique is limited since the RSSI measurements contain noise on the order of several meters due to the effects of shadowing and multipath. Another major challenge is in estimating the propagation model parameters, and the variability of the path loss exponent depending on the environment [1].

5.2 Nodes' position estimation

Having enough information (distance and/or angles), a node can compute its position using one of the nodes' position estimation techniques, such as trilateration/multilateration, triangulation, and bounding box.

5.2.1 Trilateration/multilateration

The trilateration technique is the most basic technique. Using the positions of three neighbor anchors, and the distances separating them from these three nodes, the unknown node estimates its position (**Figure 2a**).

In fact, a node must be positioned someplace along the periphery of a circle centered at the anchor's position with a radius equal to the distance separating sensor and anchor [3]. Hence, the node's position is estimated using the intersection of three circles formed by the anchors' positions and anchor-node distances. A simple system of three equations is built to compute (x, y) the unknown node's position.

5.2.1.1 Atomic multilateration

On the other hand, when the number of reference points (anchors) is more than three (n), the multilateration technique is called atomic multilateration. However, the

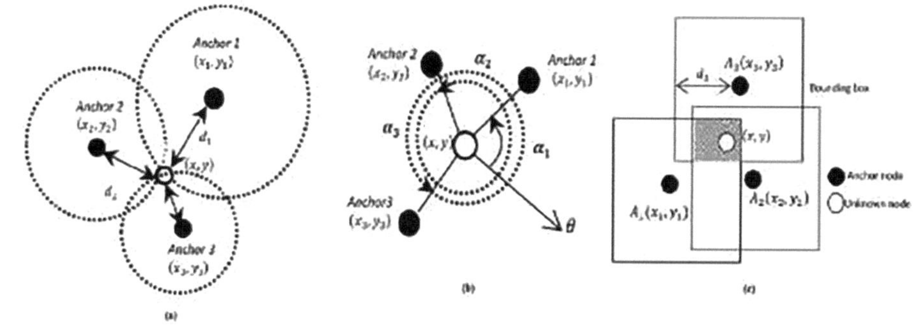

Figure 2.
Estimation nodes' position: (a) trilateration, (b) triangulation, and (c) bounding box method.

system of equations presented by Eq. (13) is an overdetermined one and it does not have a unique solution. Assuming n anchors of locations (x_i, y_i) $(i = 1, 2, \ldots, n)$, the system of equations is represented as:

$$\begin{cases} (x_1 - x)^2 + (y_1 - y)^2 = d_1^2 \\ (x_2 - x)^2 + (y_2 - y)^2 = d_2^2 \\ \vdots \\ (x_n - x)^2 + (y_n - y)^2 = d_n^2 \end{cases} \tag{13}$$

By making some arrangements, the relation above yields:

$$A\mathbf{x} = b \tag{14}$$

with

$$A = \begin{bmatrix} 2(x_n - x_1) & 2(y_n - y_1) \\ 2(x_n - x_2) & 2(y_n - y_2) \\ \vdots & \vdots \\ 2(x_n - x_{n-1}) & 2(y_n - y_{n-1}) \end{bmatrix} \tag{15}$$

$$b = \begin{bmatrix} d_1^2 - d_n^2 - x_1^2 - y_1^2 + x_n^2 + y_n^2 \\ d_2^2 - d_n^2 - x_2^2 - y_2^2 + x_n^2 + y_n^2 \\ \vdots \\ d_{n-1}^2 - d_n^2 - x_{n-1}^2 - y_{n-1}^2 + x_n^2 + y_n^2 \end{bmatrix} \tag{16}$$

This linear system is solved easily (least square approach) as shown below:

$$\mathbf{x} = \left(A^{\mathrm{T}}A\right)^{-1}A^{\mathrm{T}}b \tag{17}$$

However, assuming perfect measurements, this technique fails when the distance measurements are noisy.

5.2.1.2 Iterative and collaborative multilateration

Once a node has estimated its position, it becomes an anchor and broadcasts messages containing its estimated position to other nearby nodes. This process called iterative multilateration repeats until all nodes have been localized [21]. However, this technique accumulates localization error with each iteration.

Although it is possible that a node does not have three neighboring anchor nodes. Hence, a method called collaborative multilateration is used [22] by using location information obtained over multiple hops. This is done by constructing a graph of participating nodes that are anchors or have at least three participating neighbors. It estimates its position by solving a corresponding system of quadratic equations relating the distances between the node and its neighbors [3] as presented in Eq. (18).

$$f(x,y) = d_i - \sqrt{(x_i - x)^2 + (y_i - y)^2} \tag{18}$$

where $X = (x, y)$ is the position of the unknown node, (x_i, y_i) is the positions of anchor i, and d_i is the estimated distance between an unknown node and anchor i.

5.2.2 Triangulation

Based on information on angles instead of distance, the triangulation technique is used to determine the position of a node by using the geometric properties of triangles and trigonometric laws (**Figure 2b**). The distance to the anchor nodes is estimated using the AoA measurements.

Let $\alpha = [\alpha_1, \ldots, \alpha_n]^T$. The bearing measurements from n anchor nodes. In fact, due to Gaussian noise, $\delta\theta$, with zero mean the relationship between measured and actual bearing is:

$$\alpha = \theta(X) + \delta\theta \tag{19}$$

with $\theta(X) = [\theta_1(X), \ldots, \theta_n(X)]^T$ actual bearings.

The relationship between the bearings of anchors and their locations can be expressed in Eq. (20).

$$\tan\theta_i(x) = \frac{y_i - y}{x_i - x} \tag{20}$$

To estimate a sensor's location, different statistical methods can be applied, such as the maximum likelihood (ML) estimator.

5.2.3 Bounding box

Another type of position estimation is called bounding box (min-max algorithm). It was proposed in [23], it uses squares, instead of circles as in trilateration, to bound the possible positions of a node. This is done by constructing a bounding box for each anchor using its position and estimated distance, then determining the intersection of these boxes (**Figure 2c**). By taking the maximum of the low coordinates and the minimum of high coordinates of all bounding boxes, the shaded area (**Figure 2c**) is expressed by Eq. (21).

$$(\hat{x},\hat{y}) = \left[\max\left(x_i - d_i\right),\ \max\left(y_i - d_i\right)\right] \times \left[\min\left(x_i + d_i\right),\ \min\left(y_i + d_i\right)\right] \quad (21)$$

The final position of the unknown node is then computed as the center of the intersection of all bounding boxes [24, 25].

5.3 Range-based protocols

5.3.1 Ad hoc localization system

Using either RSS or ToA measurement, the ad hoc localization system (AHLoS) provides localization service [21]. Ranging measurements are executed by each node and then position estimation technique (discussed in Section 5.2) is used to estimate the location of unknown nodes in the network [21].

In fact, AHLoS aims to provide a distributed localization in WSNs. Moreover, it does not rely on a single type of ranging technique [1]. Localization accuracy in the range of tens of centimeters in small-scale networks is obtained while using iterative multilateration. However, in large-scale networks, using iterative multilateration leads to inaccurate results since the initial estimation errors are propagated through the net [1].

5.3.2 Time-based positioning scheme

The time-based positioning scheme (TPS) is a distributed range-based protocol, which exploits the TDoA ranging technique [26]. Three noncollinear powerful anchor nodes are deployed around the sensor network and can reach all the nodes in the network. Each anchor node periodically broadcasts a beacon message. An unknown node receives messages from anchors and tries to estimate its location through TDoA measurements. Likewise, it consists of two steps: range detection (TDoA measurements) and location computation, where trilateration is used [1]. The accuracy of this localization scheme depends on the accuracy of the TDoA measurements. Anchors are not required to be synchronized and the beacon messages can be transmitted at different times. However, the requirement of having powerful anchor nodes is not always valid for WSN architectures.

5.3.3 Mobile-assisted localization

Uniform network deployment is not feasible in practice; hence, the localization accuracy is basically limited depending on the network topology and the deployment strategy. Mobile-assisted localization (MAL) [27] uses mobile agents to improve the localization accuracy in WSNs [28, 29], it travels throughout the network to estimate the distance between sensor nodes and itself as well as between these sensors. MAL localizes the nodes using multilateration techniques.

Sensor nodes do not need to perform additional distance measurements or solve complex localization equations. However, the mobile agent required to perform localization tasks may not be available for most applications making the MAL limited in many applications [1]. In addition, the performance is highly dependent on the measurements performed at each single node; hence, any errors in the measuring

Method	Pros	Cons
DV-Hop	• Simple implementation.	• Low accuracy in a non-uniform sensor's distribution.
APIT	• Low complexity. • Applicable to scenarios where high localization accuracy is not required.	• Low accuracy. • Accuracy proportional to the number of anchors.
Centroid	• Simple and basic method.	• Heavily affected by the number of anchors.
MDS	• Ability to locate the positions of more than one node simultaneously. • Ability to have the network topology diagram in the absence of anchors.	• High traffic. • High consumption of energy. • Low accuracy in large networks.

Table 2.
Range-free algorithms.

algorithm affect the whole network. Moreover, in large networks, the localization latency may be meaningfully large because only one mobile agent is used making the time to navigate the whole network long time [1].

6. Range-free localization

Range-free techniques estimate location based techniques since they do not require additional hardware. **Table 2** summarizes some range-free algorithms as well as their pros and cons. Similar to the range-based protocols, anchor nodes may also be used to provide a reference for localization.

6.1 DV-Hop

DV-Hop localization scheme proposed by Niculescu and Nath [30] is similar to the traditional routing schemes based on the distance vector. The algorithm can be described in three steps. First, each anchor node floods a beacon message including its position and an initial value of hop field equal to zero. Neighbor nodes receive beacons and record the minimum hops to each anchor node and ignore the message with larger hops from the same anchor node [31].

Then, beacons are flooded again to their neighbor nodes with one hop increased. At the end of this step, each node in the network will eventually be able to compute the shortest path distance (in terms of hop count) from any anchor in the network [32]. When an anchor node obtains hop counts to other anchors, it estimates an average distance for one hop, which is subsequently flooded to the entire network [31]. In the second step, after obtaining hop counts to other anchors, each anchor calculates the average one-hop size, called the correction factor (e.g., anchor i).

$$\text{Hopsize}_i = \frac{\sum_{i \neq j} \sqrt{\left(x_i - x_j\right)^2 + \left(y_i - y_j\right)^2}}{\sum_{i \neq j} h_{i,j}} \tag{22}$$

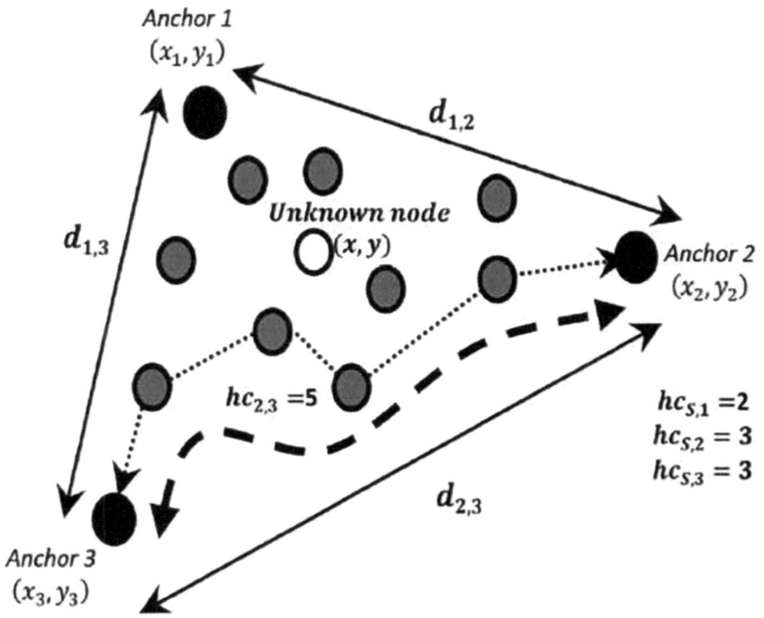

Figure 3.
DV-Hop.

where (x_i, y_i) and (x_j, y_j) are the coordinates of anchor i and anchor j, respectively, h_{ij} is the number of hops between anchor i and anchor j, each anchor node broadcasts its hop-size to network by using controlled flooding. Unknown nodes receive the information of hop-size and preserve the one received from the nearest anchor. Simultaneously, they transmit the hop-size to their neighbor nodes. After all, unknown nodes have received the hop-size from anchor nodes; they compute their distances to the anchor nodes (**Figure 3**), with h_c the minimum hop count as:

$$d_i = h_c \times \text{hopsize}_i \tag{23}$$

Finally, after receiving three or more distance information, unknown nodes estimate their positions using multilateration or ML estimation [27]. The distribution of sensor nodes plays a role in the accuracy of the DV-Hop algorithm, that is, if the inter-node distances are nearly equal, the estimated average hop-size will be accurate resulting in a low localization error. However, if the node distribution is uneven, the algorithm's accuracy is poor [33]. Many improvements are proposed in the literature to reduce errors introduced in the average hop distance calculation and multilateration. An improved DV-Hop is presented in [34], where only the third phase of DV-Hop was altered by making the unknown sensors that need to trilaterate themselves use, the 2D hyperbolic trilateration. Simulation results showed that the proposed algorithm can improve location accuracy and coverage than the original DV-Hop algorithm. Also, it showed that the more regular placement of anchors, the lower the error and higher location coverage. Work in [35] designed as well an improved DV-Hop localization algorithm, which can satisfy the node randomly distribution and heterogeneous network. In the nonuniform network distribution, the algorithm uses the weighted average to reduce the location error; the less hop-count gives great weight and the more hop-count gives small weights. Simulation and experiment results showed that the improved DV-Hop has higher positioning accuracy. A various average hop distance algorithm VAH-DV-hop is proposed in [36], it can reduce power consumption and omit any extra hardware. The principle behind this algorithm is using the angle method to reduce the harm caused by routing void (qualified anchors would execute the calculation) and applying various average hop distances (AHD) to improve the accuracy of distance estimation. The simulation results showed that VAH-DV-Hop can apparently improve the positioning accuracy, especially in uneven networks. In DV-Hop method, straight-line hop distance is substituted by hop distance. However, the path between the anchor node and unknown nodes is not a straight line in a practical network. Authors of [37] improved the accuracy of DV-Hop method by adding a correction to the distance between the anchor nodes and unknown nodes, to reduce localization errors introduced by DV-Hop.

The work in [38] proposes to use a threshold value of distance or hop count to optimize the calculation, and that to protect the dying nodes from energy drain. Moreover, authors in [39] present an algorithm to select a reasonable maximum hop count by hop-size comparison, and that by using a single-hop average error function and a sub-error estimation function to adjust the average hop distance from the source node. The error generated by using all anchors is reduced; nevertheless, the process attains a high amount of online and offline calculation.

Multilateration usually causes errors in the last phase of DV-hop. Hence, a differential evolution (DE) algorithm to rectify the accuracy is proposed in [40]. The DE algorithm uses stochastic search, which demands a highly complex operation.

The inverse distance weighting (IDW) correction method to obtain a more accurate average hop distance is applied in [41]. It is conducted by giving different weights to anchors based on the distances. In fact, the nearby anchors are assigned high weights and further away ones with lighter weights.

6.2 Approximate point-in-triangulation protocol

The approximate point-in-triangulation (APIT) protocol proposed in [42] relies on a network that consists of wireless sensor nodes as well as anchors. It consists of four phases:

1. Beacon exchange: Each node is informed about the connectivity of each of its neighbors to the anchors, and it builds up a table and broadcasts it to its neighbors.

2. PIT test: A node is determined to be inside/outside a triangle formed by three anchor nodes. It is considered outside the triangle if the distances to the vertexes of the triangle increase or decrease simultaneously when it moves along any direction. Otherwise, it is considered inside the triangle.

3. APIT aggregation: This test determines the triangles in which the unknown node exists. Then, an aggregation is performed to constrain the location and that by calculating the maximum overlapping area of these triangles.

4. COG: The node estimates its location as the center of gravity of the overlapping area.

The complexity of the protocol is significantly low compared to range-based algorithms. However, the localization accuracy of the algorithm is also low and is proportional to the number of nonlinear anchors connected to unknown nodes. Consequently, the APIT protocol is applicable to scenarios where high localization accuracy is not required [1].

An algorithm based on energy threshold (ET-APIT) is proposed in [43] to reduce the probabilities of In-to-Out error and Out-to-In error in APIT localization algorithm. By introducing a certain energy threshold, the unknown nodes that are too close to the anchor node triangles causing an error in estimation are removed, and by using iteration, the located unknown nodes are seen as anchor nodes to locate more unknown nodes.

Moreover, when the location of the unknown node is near the edge, the accuracy of the APIT is low; hence, a work in [44] is proposed to eliminate the edge error effect by applying the Barycentric Coordinate Technique.

6.3 Multidimensional scaling

Multidimensional scaling (MDS) is a technique that has its origins in psychometrics and psychophysics [45]. MDS technique, applied to solve localization problems in WSN, displays the structure of distance-like data as a geometrical picture [3]. Its goal is to implement a projection technique capable of preserving the similarities present in the original data set. Hence, the network can be recreated in the multidimensional space. As a result of MDS algorithm, the network layout will be an arbitrarily rotated and flipped version of the original one [3].

The MDS map [46] is a proposed localization method based on MDS technique, which provides both relative and absolute maps. It uses the connectivity information to derive the location of the nodes in the network. Initially, using inter-node distances of all nodes, it constructs the relative map. Then, using enough anchors, it can estimate absolute coordinates by transforming relative maps into absolute map. Since MDS-MAP uses the length of the shortest path as Euclidian distance between the nodes, it is sensitive to the shape of the network. Thus, it presents poor performance on irregular networks, since the difference between shortest path distance and actual Euclidian distance causes large error [47]. An improvement of MDS-MAP is presented in [47], where the last step is different than the original approach in order to solve the problem of irregular networks by dividing the irregular network into several sub-networks. For each sub-network, distribution of the nodes is relatively uniform. Therefore, individual linear transformation can be employed to separately map the coordinates of each set of nodes from the relative map to their absolute coordinates.

Classical MDS localization algorithm has a low accuracy in large-scale sensor network with a lot of nodes. For this defect, the authors of [48] proposed an improved algorithm based on fuzzy-c means. This is done by splitting the network by using a fuzzy c-means clustering algorithm and then applying an MDS localization algorithm in sub-networks.

6.4 Centroid

One of the simplest solutions in range-free localization is the centroid [49]. Its scheme is mainly based on anchors. All anchors send their positions to all nodes within their communication radius. These latter determine their locations by computing the average value of the anchor coordinates heard, that is, the center of gravity, of a system of masses placed in correspondence of the anchor nodes heard [32] calculated as:

$$\left(x = \frac{\sum_{i=1}^{M} x_i}{M}, y = \frac{\sum_{i=1}^{M} y_i}{M}\right) \tag{24}$$

with (x, y) the unknown node's location, (x_i, y_i) the anchor's i location, and M the number of anchors. The centroid localization algorithm is simple, but it is heavily affected by the number of anchor nodes used. It fits for high anchor density homogenous networks.

6.4.1 Improved centroid: Weighted centroid

The work in [50] proposed a weighted centroid algorithm. The reference node is the nearest to the unknown node. Also, the nodes localized (position estimated) are called upgrade anchor nodes. We summarize this algorithm in 5 steps:

1. The weight is calculated based on the distance between the reference anchor node and other anchor nodes.

2. A number of triangles are formed between the reference anchor node and other anchor nodes.

3. The centroid of these triangles is calculated, then the weight value calculated above is used to weight the group's centroid, and then calculate the weighted centroid.

4. Finally, the node is localized and upgraded to anchor nodes.

5. This algorithm is applied to all unknown remaining unknown nodes.

7. Centralized versus distributed paradigm

In a centralized algorithm, all sensor nodes in the network send their data to the central receiver and receive their computed locations. It requires plenty of computational power in order to run their operations on central machines enabling the algorithms to execute complex mathematical operations (order of $O(n^2)$ and $O(n^3)$) [51], which results in a high precision localization, high energy consumption, and a robust scaling effect. It requires that a powerful base station can be deployed among the nodes. However, this process leads to a high communication cost.

On the other hand, in a distributed algorithm, operations are processed using the computational power of each node. Thus, massive inter-node communication and parallelism are required to be able to perform similar to centralized systems [51]. Besides, it is a low-energy consumer, and a robust algorithm when scaling. However, it presents a limited precision due to noncomplex mathematical operations used.

8. Mobile versus static sensors

According to the application and the field of sensor nodes in which they are deployed, sensor nodes are either static and fixed at one place or mobile. A WSN is considered mobile when nodes can move and leave their position to another one, hence, WSN topology changes. Localization in this case is performed to track them, or for navigational purposes. In fact, four combinations of mobility can be discussed:

1. Static sensor nodes and static anchor nodes

2. Static sensor nodes and mobile anchor nodes

3. Mobile sensor nodes and static anchor nodes

4. Mobile sensor nodes and mobile anchor nodes

Three categories discuss the mobility in a WSN:

1. Random mobility: where the sensors move randomly in the area of deployment.

2. Predictable mobility: where the motion of sensors is known but cannot be changed.

3. Controlled mobility: where the sensors move to definite destinations following defined mobility outlines.

Many mobility models are proposed to describe a node's movement, such as Random Way (RW) [52], Random WayPoint mobility (RWP) model [53], Gauss-Markov (GM) [54], and Boundless Mobility model [55].

In fact, localization techniques can vary the anchor node density. Hence, mobile anchor nodes collaborate with the static sensor nodes to make up the constraint of localization in static WSNs. Work done in [56] reviewed most MANAL (Mobile Anchor Node Assisted Localization) algorithms. It divides the movement trajectories into two types: the first where the anchors move with some already existing mobility models without considering network parameters and localization, and the second one where they move with some path scheduling outlines designed for WSN localization. However, when sensors move additional challenges are encountered such as localization latency. If the time to estimate the position of the node is too long, the sensor will have changed its position. Also, mobility may impact the localization signal; the frequency of the signal may experience a Doppler shift which occurs when the transmitter of a signal is moving relative to the receiver. This shift in frequency is correlated to the positions of the two nodes [57]. Work in [58] took this Doppler effect into account and uses it to improve the estimated position.

Sensor's mobility causes distance variations and environmental interference. However, a well-designed localization technique can reduce the number of reference anchors required. Also, the network performance is enhanced in terms of packet delay, coverage (better deployment) [59], and connectivity [60]. Moreover, the communication overhead is reduced as well as the energy consumption, which increases the durability of the whole network. However, the localization estimation error is a function of the speed of the anchor nodes and sensor nodes.

9. Fingerprinting technique

Another category of localization techniques is fingerprinting technique or scene analysis. It uses the signatures or fingerprints and is based on a study campaign conducted in the environment where the location system is deployed. It consists of two phases: the off-line phase where a signature database is built and the real-time phase where the location of the node is estimated by comparing the current signature with those cataloged previously. Several types of signatures [61] can be used: the power of the received signal, the AoA, the arrival time, the delay spread, or the number of reflected paths of received signals. A pattern-matching algorithm is used such as K-Nearest-Neighbor, KNN [62], Kernel-based [63], histogram method, support vector machines (SVM) [64], smallest M-vertex polygon (SMP), random forest [65], decision trees [66], and artificial neural networks [67]. Database building is a relatively simple process: (1) It does not require the receiver to connect to the transmitter and exchange messages. (2) It is not necessary to know the transmitters' position information. However, this technique suffers from noise, and any change in the environment decreases localization accuracy. However, the requirement for generating a signal signature database makes this technique unachievable for the most scenarios of the WSNs, especially in complex environments.

The level of obtained accuracy depends on how many access points and reference points are used. Localization accuracy is enhanced with access points number, also, the resolution is enhanced with reference nodes number; however, this will cost more labor work. Another known drawback of this approach is the need for regular updates for the collected data as well as the built map [68].

10. Three-dimensional localization aspect additional challenges

The majority of localization techniques have been proposed considering only two-dimensional (2D) networks. Henceforth, localization in 3D is an interesting problem in the research community. Landscape-3D [69] is one of the first proposed techniques for 3D localization, where unknown nodes measure a set of distances to mobile location assistants (LAs) using RSSI, then they use unscented Kalman filter to estimate their own position. Also, in [70], RSSI is used for distance measurements while particle filter is used for node positioning. On the other hand, an improved centroid localization method is presented in [71], where each unknown node randomly chooses four anchor nodes in range to form a sequence of tetrahedrons used to calculate its position. In [72], a range-free algorithm is proposed based on flying anchors. In fact, mobile anchor nodes keep transmitting a beacon message along with their location information to unknown nodes and choose three further anchor nodes to form a triangle. Then, the distance is calculated by the link quality induction against each anchor node. Finally, a centroid algorithm is used to estimate the node's position.

However, some difficulties are faced in 3D localization algorithms [73] such as:

- More anchor nodes are needed for localization; in fact, at least three anchor nodes are required in a 2D space, whereas, in a 3D space, it needs at least four anchor nodes to locate the unknown nodes. Hence, the node density increases as well as the complexity of the algorithm.

- Transmitted signals are affected by the terrain obstacles, affecting the distance estimation between nodes, which will affect the positioning accuracy.

11. Fundamental limitations impacting localization

11.1 Number of anchors

It has been shown that the localization accuracy increases with the number of anchors [74, 75]. Nevertheless, in some scenarios the number of available anchors is low for different reasons such as battery exhaustion or limited communication range [76]. Hence, the localization is limited in these cases.

11.2 Distribution of anchors

The distribution and deployment of anchor nodes play an important role in the localization algorithm. If the anchors are placed only in some portion of the area of interest, it does not guarantee that all unknown nodes reside inside the convex hull formed by the anchors resulting in a low localization accuracy [76]. The geometric dilution of precision (GDOP) is a parameter used to interpret the relation between anchor distribution and accuracy which increases as the value of GDOP decreases [77]. The GDOP is used in optimizing the deployment of the sensors.

11.3 Nonline of sight

The nonline of sight is defined when the propagation path, between the transmitter and receivers, is obstructed. Hence, the communication between nodes may be lost, limiting the localization accuracy. The effects of this phenomenon are more important when the elements in the environment are regularly changing. If there exists information on the NLOS links, it can be used to improve the localization accuracy [76].

11.4 Multipath propagation

Multipath propagation occurs when the transmitted signal arrives at the receiver by two or more paths. It causes constructive and destructive interferences, altering the signal-related measurements and hence affecting the localization accuracy. For example, in the RSS-based localization; the transmitter sensor seems to be farther away than where it is in reality. An Optimal Multi-Channel Trilateration positioning algorithm (OMCT) is presented in [78]. It first uses an adaptive Kalman filter to remove the RSS measurement noise and the optimal node position estimates are obtained from a multiobjective evolutionary algorithm.

12. Localization performance indicator

12.1 Accuracy of localization

The error of localization defined as the Euclidean distance between the real and estimated positions of nodes is the most important feature in localization evaluation.

To increase the accuracy of the localization, algorithm has to minimize this error. However, factors affecting the hardware, the processor, and the energy (such as size and cost) must be taken into consideration.

12.2 Complexity

A localization algorithm must be fast, noncomplex, and its development does not require large calculations and large memory storage capacity. For instance, if the complexity is the major property to take into consideration in a localization algorithm, the trilateration method is suitable; however, it is susceptible to inaccurate distances' estimations.

12.3 Energy constraints

The only energy source of a sensor node is its battery. Hence, careful energy management is required in a WSN to avoid wasting it, it is necessary that the algorithm communicates the least possible via radio. Schemes based on hop-count require high communication cost. Thus, the localization scheme should minimize the amount of node-to-node communication.

12.4 Scalability

Localization technique must ensure appropriate estimation of position when WSN deployment gets larger. In fact, when the distance between nodes increases, the performance of range-based techniques decreases. Moreover, in dense network signals are subject to congestion requiring complex infrastructure.

13. Research directions and challenges

In order to obtain more accurate and better performance of localization algorithms, multimodal localization is more investigated, where, multiple localization techniques are used simultaneously. Work in [79] exploited a hybrid TOA/RSS range estimator combined with an iterative least-squares procedure to localize nodes. The proposed hybrid approach outperformed state-of-the-art techniques. Another hybrid approach is proposed in [80], where a localization based on TOA/AOA techniques is presented. Elevation AoA estimations are combined with ToA measurements, then applied to a weighted least square algorithm to solve the nonlinear problem. Simulation results show that the proposed method outperforms the conventional methods, by adjusting different parameters such as transmit power, signal bandwidth, and the number of anchors. Authors in [81] proposed an approach using hybrid RSS and AOA to resolve a source localization problem in a 3D WSN. RSS model integrates the Gaussian-shaped radiation pattern, and the technique adopts the second-order cone relaxation and alternating optimization techniques. Simulation results demonstrate the efficacy of the presented algorithm.

Another aspect of research directions is the heterogenous WSN. The work in [82] proposed a fault filtering method used with an existing hop-based algorithm. First, it normalizes the distance estimations using the communication radius of nodes and then uses the Jenks Natural Breaks algorithm for filtering out the nodes producing unreliable distance estimations. The approach is tested in 2D/3D, isotropic/anisotropic networks. Localization accuracy shows an improvement of 14 and 52% when tested with DV-Hop, Weighted DV-Hop. Another approach [83] is a priority-based algorithm, which gives priority to a few anchors based on their AHD. Unknown nodes are then localized with weighted centroid method using high priority anchors. Results show that algorithm outperforms existing weighted centroid methods in anisotropic fields.

An additional research direction considered localization in irregular field. Irregularities present challenges in nodes localization, and they can be signified in terms of irregular radio propagation pattern of nodes, noisy environment, network holes, and irregular fields [84]. It is useful and important in environmental applications such as forest fire monitoring, however, forest areas are usually not plain uniform fields. Hence, considering irregularities increase localization accuracy [85]. In fact, RSS-based localization techniques are affected by irregularities, since RSS values between a pair of transmitters and receivers at fixed distance varied when the receiver was placed at different propagation directions from the transmitter [86]. Hence, a novel technique where node segmentation with improved particle swarm optimization (NS-IPSO) is proposed in [87]. It divides sensor nodes into segments to improve the accuracy of the estimated distances between pairs of anchor nodes and unknown nodes. Similarly, irregularities add a positive bias for the TOA and TDOA measurements [88], resulting in overestimation of distance between nodes and higher localization errors. A neural network-based localization algorithm called LPSONN was described in [89], it is a centralized algorithm implemented and simulated in isotropic networks with and without coverage holes or shadowing zones, and anisotropic networks. A neural network using the received information is trained. Results show that the proposed algorithm has less localization error rate and storage requirements than the analogous methods.

14. Future scopes

The evolution of WSN, technologies, as well as localization applications create the necessity of more advanced research exploiting intelligent surfaces as well as advanced millimeter-wave systems. Future scopes and studies are concerned by a new concept that emerged recently called Reconfigurable Intelligent Surfaces (RISs). In fact, future WSN will not only allow people and devices localization but will be turned into a distributed intelligent communication, sensing, and computing platform [90].

RIS may be able to propose dense networks for sensing the environment and to offer a platform that provides highly accurate localization services in outdoor and indoor scenarios, by taking advantages of realizing large-size smart surfaces. Also, RISs can offer a possibility to acquire a fully electromagnetic-based computing platform and that thanks to the possibility of performing algebraic operations and functions directly on the incident radio waves [90]. In addition, RIS can present important advantages in terms of performance, energy consumption, and cost for localization and mapping [91].

Besides, systems where antenna arrays, are deployed as a large intelligent surface (LIS) are a prospective field for positioning and coverage enlargement of wireless networks [92].

More interesting future scopes and studies will be based on the joint usage of RISs and millimeter wave MIMO systems for the fifth generation (5G) [93], where evaluation of the impact of the number of LIS elements are studied and the theoretical performance for localizing are compared to the conventional scheme with one direct link and one non-line-of-sight path [93].

Hence, several researchers have started investigating several scopes and opportunities offered by RIS as well as the envisioned 6G platform, which is expected to sense the environment, store and process information to provide network applications.

15. Conclusion

Localization in WSN is an important and challenging task, it is essential for many applications and network management. This chapter surveys the most popular range-based and range-free techniques. It presents the basics of each one as well as the research directions. Readers can profit from this chapter to well understand the concepts of localization in WSN. Different works are summarized in this work, allowing readers and researchers to be positioned with respect to enhancements and ideas presented in the literature.

Nevertheless, localization and mapping algorithms discussed and detailed in this chapter can benefit from using RIS facilities, in which position and orientation are known a priori [91], improving, hence, the accuracy and extending radio coverage.

Moreover, this concept has high potential approaches for next-generation localization, and more importantly when investigations consider beyond 5G localization.

Author details

Nour Zaarour*, Nadir Hakem and Nahi Kandil
University of Quebec in Abitibi-Temiscamingue-Research Laboratory Télébec in Underground Communications (UQAT-LRTCS), School of Engineering, Val-d'Or, Quebec, Canada

*Address all correspondence to: nour.zaarour@uqat.ca

References

[1] Akyilidz IF, Vuran MC. Wireless Sensor Networks. United Kingdom: John Wiley & Sons Ltd; 2010. p. 520

[2] What is the Internet of Things? WIRED Explains [Internet]. Available from: https://www.wired.co.uk/article/internet-of-things-what-is-explained-iot

[3] Dargie W, Poellabauer C. Fundamentals of Wireless Sensor Networks Theory and Practice. United Kingdom: John Wiley & Sons Ltd; 2010. p. 336

[4] Viani F, Rocca P, Oliveri G, Trinchero D, Massa A. Localization, tracking, and imaging of targets in wireless sensor networks: An invited review. Radio Science. 2011;**46**(05):1-12. DOI: 10.1029/2010RS004561

[5] Dovis F, Margaria D, Mulassano P, Dominici F. Overview of Global Positioning Systems. In Handbook of Position Location: Theory, Practice, and Advances, 2nd ed.; Zekavat R, Buehrer RM, Eds.; Wiley-IEEE Press: Hoboken, NJ, USA;2018, Chapter 20; pp. 655–705. DOI:10.1002/9781119434610

[6] Kandris D, Nakas C, Vomvas D, Koulouras G. Applications of wireless sensor networks: An up-to-date survey. Applied System Innovation. 2020;**3**(1): 14. DOI: 10.3390/asi3010014

[7] Hii P, Chung W. A comprehensive ubiquitous healthcare solution on an Android™ mobile device. Sensors. 2011; **11**:6799-6815

[8] Giorgetti G. Resource-constrained localization in sensor networks [Ph.D. dissertation]. University of Florence, Italy: Department of Electronics and Telecom; 2007.

[9] Niu R, Vempaty A, Varshney PK. Received-signal-strength-based localization in wireless sensor networks. Proceedings of the IEEE. 2018;**106**(7): 1166-1182. DOI: 10.1109/JPROC.2018.2828858

[10] Chuku N, Nasipuri A. RSSI-based localization schemes for wireless sensor networks using outlier detection. Journal of Sensor and Actuator Networks. 2021;**10**(1):10. DOI: 10.3390/jsan10010010

[11] Meghani S K, Asif M, Amir S. Localization of WSN node based on time of arrival using ultra wide band spectrum. In: WAMICON 2012 IEEE Wireless & Microwave Technology Conference; 15-17 April 2012; Cocoa Beach, FL, USA: IEEE; 2012. pp. 1-4. DOI:10.1109/WAMICON.2012.6208430

[12] Feng R, Li C, Ran Q, Wu Y, Yu N. A novel TOA-based source localization algorithm in wireless sensor networks. In: Eighth International Conference on Information Science and Technology (ICIST); 30 June-6 July 2018. Cordoba, Granada, and Seville, Spain: IEEE; 2018. pp. 429-436. DOI: 10.1109/ICIST.2018.8426135

[13] Meng W, Xie L, Xiao W. Decentralized TDOA sensor pairing in multihop wireless. IEEE Signal Processing Letters. 2013;**20**(2):181-184. DOI: 10.1109/LSP.2013.2237823

[14] Wang T, Xiong H, Ding H, Zheng L. TDOA-based joint synchronization and localization algorithm for asynchronous wireless sensor networks. IEEE Transactions on Communications. 2020; **68**(5):3107-3124. DOI: 10.1109/TCOMM.2020.2973961

[15] Shao H, Zhang X, Wang Z. Efficient closed-form algorithms for AOA based self-localization of sensor nodes using auxiliary variables. IEEE Transactions on Signal Processing. 2014;**62**(10): 2580-2594. DOI: 10.1109/TSP.2014.2314064

[16] Arbula D, Ljubic S. Indoor localization based on infrared angle of arrival sensor network. Sensors. 2020; **20**(21):6278. DOI: 10.3390/s20216278

[17] Ghelichi A, Yelamarthi K, Abdelgawad A. Target localization in wireless sensor network based on time difference of arrival. In: Proceedings of IEEE 56th International Midwest Symposium on MWSCAS; 4-7August 2013. Columbus, OH: IEEE; 2013. pp. 940-943. DOI: 10.1109/MWSCAS.2013.6674805

[18] Cao H, Wireless CX. Sensor Networks: Principles and Practice. 1st ed. Boca Raton, Fla, USA: CRC Press; Auerbach; 2010

[19] Peng P, Sichitiu ML. Angle of arrival localization for wireless sensor networks. In: 3rd Annual IEEE Communications Society SECON '06. Reston; 28-28 September 2006. Reston, VA, USA: IEEE;2007. pp. 374-382. DOI: 10.1109/SAHCN.2006.288442

[20] Rappaport TS. Wireless Communications, Principle & Practice. Upper Saddle River, N.J: 2nd ed. Prentice Hall; 2011

[21] Savvides A, Han CC, Strivastava MB. Dynamic fine-grained localization in ad hoc networks of sensors. In: Proceedings of the 7th Annual International. Conference on Mobile Computing and Networking; July 2001. New York, USA. DOI: 10.1145/381677.381693

[22] Avanthi K, Xinrong L, Murali V. Comparative study of RSS-based collaborative localization methods in sensor networks. In: Proceedings of IEEE Region 5 Conference; 7-9 April 2006. San Antonio, TX, USA: IEEE; 2010. pp. 243-248. DOI: 10.1109/TPSD.2006.5507424

[23] Doherty L, Pister KSJ, El Ghaoui L. Convex position estimation in wireless sensor networks. In: Proceedings of IEEE 20th Annual Joint Conference of the Computer and Communications Societies INFOCOM; 22-26 April 2001. Anchorage, AK, USA: IEEE; 2002. pp. 1655-1663. DOI: 10.1109/INFCOM.2001.916662

[24] Shi X, Zhang L. High-precision weighted bounding box localization algorithm for wireless sensor network. In: Proceedings of IEEE International Conference on ICIST; 23-25 March 2013. Yangzhou: IEEE; 2014. pp. 1110-1113. DOI: 10.1109/ICIST.2013.6747730

[25] Aiping P, Xiaosong G, Wei C, Haibin L. A distributed localization scheme for wireless sensor networks based on bounding box algorithm. In: Proceedings of IEEE International Conference on ICEMI; 16-19 August. 2009. Beijing, China: IEEE; 2009. pp. 984-988

[26] Cheng X, Thaeler A, Xue G, Chen D. TPS: A time-based positioning scheme for outdoor wireless sensor networks. In: Proceedings of IEEE 23th Annual Joint Conference on Computer and Communications Societies INFOCOM; 7-11 March 2004. Vol. 4. Hong Kong, China: IEEE; 2004. pp. 2685-2696. DOI: 10.1109/INFCOM.2004.1354687

[27] Priyantha NB, Balakrishnan H, Demaine ED, Teller S. Mobile-assisted localization in wireless sensor networks. In: Proceedings of IEEE 24th Annual Joint Conference on Computer and Communications Societies INFOCOM, Volume 1; 13-17 March 2005. Miami, FL, USA: IEEE; 2005. pp. 172-183. DOI: 10.1109/INFCOM.2005.1497889

[28] Wang H, Qi W, Wang K, Liu P, Wei L, Zhu L. Mobile-assisted localization by stitching in wireless sensor networks. In: Proceedings of IEEE Conference, ICC; June 2011. Kyoto, Japan; 2011. pp. 1-5. DOI: 10.1109/icc.2011.5962799

[29] Guo Z, Guo Y, Hong F, Jin Z, He Y, Feng Y, et al. Perpendicular intersection: Locating wireless sensors with mobile beacon. IEEE Transactions on Vehicular Technology. 2010;**59**(7):3501-3509. DOI: 10.1109/TVT.2010.2049391

[30] Niculescu D, Nath B. Ad hoc positioning system (APS). In: IEEE GLOBECOM '01; November 2001. San Antonio, TX, USA. pp. 2926-2931. DOI: 10.1109/GLOCOM.2001.965964

[31] Ding J, Zhang L, Cheng G, Ling Z, Zhang Z, Lei Z. Study on DV-Hop algorithm based on modifying hop count for wireless sensor networks. International Journal of Computer Science & Engineering Technology. 2012;**2**(10):1452-1456

[32] Phoemphon S, So-In C, Leelathakul N. Optimized hop angle relativity for dv-hop localization in wireless sensor networks. IEEE Access. 2018;**6**:78149-78172. DOI: 10.1109/ACCESS.2018.2884837

[33] Gayan S, Dias D. Improved DV-Hop algorithm through anchor position re-estimation. In: Proceedings of IEEE Wireless and Mobile Asia Pacific Conference; Conference; 28-30 August 2014. Bali, Indonesia: IEEE; 2014. pp. 126-131. DOI: 10.1109/APWiMob.2014.6920272

[34] Chen H, Sezaki K, Deng P, Cheung SH. An improved DV-Hop localization algorithm for wireless sensor networks. In: Proceedings of IEEE Conference on ICIEA; 3-5 June 2008. Singapore: IEEE; 2008. pp. 1557-1561. DOI:10.1109/ICIEA.2008.4582780

[35] Chen K, Wang ZH, Lin M, Yu M. An improved DV-Hop localization algorithm for WIRELESS SENSOR NETWORKS. In: Proceedings of International Conference, IET-WSN; 15-17 November 2010. Beijing, China: IET; 2011. pp. 255-259. DOI: 10.1049/cp.2010.1063

[36] Liu J, Wang W, Shang W. An improving localization algorithm for wireless sensor networks based on DV-Hop. In: Proceedings of International Conference on MIC; 18-20 May 2012 Vol. 1. Harbin, China: IEEE; 2012. pp. 511-515. DOI: 10.1109/MIC.2012.6273353

[37] Yu W, Li H. An improved DV-Hop localization method in wireless sensor networks. In: Proceedings of IEEE International. Conference on CSAE; 25-27 May 2012. Zhangjiajie, China: IEEE; 2012. pp. 199-202. DOI: 10.1109/CSAE.2012.6272938

[38] Jin W, Yu G, Xiang Y, Li F, Kim H-J. An enhanced PEGASIS algorithm with mobile sink support for wireless sensor networks. Wireless Communications and Mobile Computing. 2018, 2018: 9472075

[39] Xiang MT, Wang S, Yang Y. Improved DV-Hop localization algorithm based on threshold mechanism and distance correction. Journal of Transistor. 2016:138-144

[40] Cui LZ, Xu C, Li GH, Ming Z, Feng YH. A high accurate localization algorithm with DV-Hop and differential evolution for wireless sensor network. Applied Soft Computing. 2018;**68**:39-52. DOI: 10.1016/j.asoc.2018.03.036

[41] Jing W, Anqi H, Yuanfei T. Hong Yu. An improved dv-hop localization algorithm based on selected anchors. Computers, Materials & Continua. 2020; **65**(1):977-991. DOI: 10.32604/cmc.2020.011003

[42] He T, Huang C, Blum B, Stankovic J, Abdelzaher T. Range-free localization schemes for large-scale sensor networks. In: Proceedings of ACM MobiCom; 14-17 September 2003. San Diego, CA, USA: ACM;2003; 2003. pp. 81-95

[43] Yang GX, Jing B. Localization algorithm for wireless sensor network based on energy threshold. In: Proceedings of International Conference on ICM; 24-25 September 2011. Vol. 2. Nanjing, Jiangsu, China: IEEE; 2011. pp. 193-197. DOI: 10.1109/ICM.2011.376

[44] Jaya PA. Analysis and implementation of APIT localization algorithm for wireless sensor network. In: 3rd International Conference on Computer and Communication Systems (ICCCS); 27-30 April 2018. Nagoya, Japan: IEEE; 2018; 2018. pp. 310-313. DOI: 10.1109/CCOMS.2018.8463241

[45] Borg I, Groenen PJF. Modern Multidimensional Scaling. New York: Springer; 1997

[46] Shang Y, Ruml W, Zhang Y, Fromherz MPJ. Localization from mere connectivity. In Proceedings of the 4th ACM International Symposium on Mobile ad Hoc Networking & Computing (MobiHoc '03). Association for Computing Machinery; 1-3 June 2003. New York, NY, USA: ACM; 2003. pp. 201–212. DOI: 10.1145/778415.778439

[47] Huang C, Xu Z, Li X. Analysis and improvement for MDS localization algorithm. In: Proceedings of the 3rd IEEE ICSESS; 22-24 June 2012. Beijing, China: IEEE; 2012. pp. 12-15. DOI: 10.1109/ICSESS.2012.6269394

[48] Stojkoska BR, Kirandziska V. Improved MDS-based algorithm for nodes localization in wireless sensor networks. In: Proceedings of IEEE EuroCON; 1-4 July 2013. Zagreb, Croatia: IEEE; 2013. pp. 608-613. DOI: 10.1109/EUROCON.2013.6625044

[49] Bulusu N, Heidemann J, Estrin D. GPS-less low-cost outdoor localization for very small devices. IEEE Personal Communications. 2000;**7**(5):28-34. DOI: 10.1109/98.878533

[50] Yang X, Wang X, Wang W. An improved centroid localization algorithm for WSN. In: 2018 IEEE 4th Information Technology and Mechatronics Engineering Conference (ITOEC); 14-16 December 2018. Chongqing, China: IEEE; 2019. pp. 1120-1123. DOI: 10.1109/ITOEC.2018.8740471

[51] Kulaib AR, Shubair RM, Al-Qutayri MA, Ng JWP. An overview of localization techniques for wireless sensor networks. In: 2011 International Conference on Innovations in Information Technology; 25-27 April 2011. Abu Dhabi, United Arab Emirates: IEEE; 2011. pp. 167-172. DOI: 10.1109/INNOVATIONS.2011.5893810

[52] Johnson DB, Maltz DA. Dynamic source routing in ad hoc wireless networks. In: Imielinski T, Korth HF, editors. Mobile Computing. The Kluwer International Series in Engineering and Computer Science. Vol. 353. Boston, MA: Springer. DOI: 10.1007/978-0-585-29603-6_5

[53] Lu G, Manson G, Belis D. Mobility modeling in mobile ad hoc networks with environment-aware. Journal of Networks. 13-15 December 2005. Wuhan, China: Academy Publisher;

2006. pp. 654-665. DOI: 10.4304/jnw.1.1.54-63

[54] Liang B, Haas ZJ. Predictive distance-based mobility management for PCS networks. In: IEEE Information Communications Conference; March 1999; New York, USA: IEEE; 2002. vol. 3. pp. 1377-1384. DOI:10.1109/INFCOM.1999.752157.

[55] Camp T, Boleng J, Davies V. A survey of mobility models for ad hoc network research. Mobile & Wireless Communications. 2002; **2002**(2):483-502. DOI: 10.1002/wcm.72

[56] Han G, Jiang J, Zhang C, Duong TQ, Guizani M, Karagiannidis GK. A survey on mobile anchor node assisted localization in wireless sensor networks. IEEE Communications Surveys & Tutorials. 2016;**18**(3):2220-2243. DOI: 10.1109/COMST.2016.2544751

[57] Amundson I, Koutsoukos XD. A survey on localization for mobile wireless sensor networks. In: Fuller R, Koutsoukos XD, editors. Mobile Entity Localization and Tracking in GPS-less Environments. MELT. Vol. 5801. Berlin, Heidelberg: Springer; 2009. Lecture Notes in Computer Science. DOI: 10.1007/978-3-642-04385-7_16

[58] Kus'y B, Sallai J, Balogh G, L'edeczi A, Protopopescu V, Tolliver J, et al. Radio interferometric tracking of mobile wireless nodes. In: Proceedings of MobiSys of 5th International Conference on Mobile Systems, Applications and Services; 11-14 June 2007. New York, NY, USA: ACM; 2007 pp. 139-151. DOI: 10.1145/1247660.1247678

[59] Kuriakose J, Amruth V, Sandesh AG, Abhilash V, Kumar GP, Nithin K. A review on mobile sensor localization. In: Proceedings of the 2nd international symposium, SSCC. Berlin, Heidelberg: Springer; 2014. pp. 30-44. DOI: 10.1007/978-3-662-44966-0_4

[60] Akcan H, Kriakov V, Bronnimann H, Delis A. Managing cohort movement of mobile sensors via GPS-free and compass-free node localization. Journal of Parallel and Distributed Computing. 2010;**70**:743-757. DOI: doi.org/10.1016/j.jpdc.2010.03.007

[61] Nerguizian C, Despins C, Affes S. A framework for indoor geolocation using an intelligent system. In: 3rd IEEE Workshop on WLANs; September 2001. Boston, USA: IEEE; 2001

[62] Fang XM, Jiang ZH, Nan L, Chen LJ. Optimal weighted K-nearest neighbor algorithm for wireless sensor network fingerprint localization in noisy environment. IET Communications. 2018;**12**(10):1171-1177. DOI: doi.org/10.1049/iet-com.2017.0515

[63] Kushki A, Plataniotis KN, Eenetsanopoulos AN. Kernel-based positioning in wireless local area networks. IEEE Transactions on Mobile Computing. 2007;**6**(6):689-705

[64] Khatab ZE, Moghtadaiee V, Ghorashi SA. A fingerprint-based technique for indoor localization using fuzzy least squares support vector machine. In: Iranian Conference on Electrical Engineering (ICEE); 2-4 May 2017. Tehran, Iran: IEEE; 2017. pp. 1944-1949. DOI: 10.1109/IranianCEE.2017.7985373

[65] Wang Y, Xiu C, Zhang X, Yang D. WiFi indoor localization with CSI fingerprinting based random forest. Sensors. 2018;**18**(9):2869. DOI: doi.org/10.3390/s18092869

[66] Huang P, Zhao H, Liu W, Jiang D. MAPS: Indoor localization algorithm

based on multiple AP selection. Mobile Networks Applications. 2021;26: 649–656. DOI: 10.1007/s11036-019-01411-7

[67] Gucciardo M, Tinnirello I, Dell'Aera GM, Caretti M. A flexible 4G/5G control platform for fingerprint-based indoor localization. In: IEEE INFOCOM 2019-IEEE Conference on Computer Communications Workshops (INFOCOM WKSHPS); 29 April-2 May 2019. Paris, France: IEEE; 2019. pp. 744-749. DOI: 10.1109/INFCOMW.2019.8845272

[68] Obeidat H, Shuaieb W, Obeidat O, Abd-Alhameed R. A review of indoor localization techniques and wireless technologies. Wireless Personal Communications. 2021;**119**:289-327. DOI: doi.org/10.1007/s11277-021-08209-5

[69] Zhou X, Zhang L, Cheng Q. Landscape-3D: A robust localization scheme for sensor networks over complex 3D terrains. In: Proceedings of the 31st Annual IEEE Conference on Local Computer Networks (LCN '06); November 2006. Tampa, FL, USA: IEEE; 2007. pp. 239-224. DOI: 10.1109/LCN.2006.322106

[70] Caballero F, Merino L, Maza I, Ollero A. A particle filtering method for wireless sensor network localization with an aerial robot beacon. In: Proceeding of the IEEE International Conference on Robotics and Automation (ICRA '08); 19-23 May 2008. Pasadena, CA, USA: IEEE; 2008. pp. 596-601. DOI: 10.1109/ROBOT.2008.4543271

[71] Chen H, Huang P, Martins M, So HC, Sezaki K. Novel centroid localization algorithm for three-dimensional wireless sensor networks. In: Proceedings of the International Conference on Wireless Communications, Networking and Mobile Computing (WiCOM '08); October 2008. Dalian, China: IEEE; 2008. pp. 596-601

[72] Javed I, Tang X, Shaukat K, Sarwar MU, Alam TM, Hameed I, et al. V2X-based mobile localization in 3D wireless sensor network. Security and Communication Networks. 2021;**2021**: 6677896. DOI: 10.1155/2021/6677896

[73] Xu Y, Zhuang Y, Gu J. An improved 3D localization algorithm for the wireless sensor network. International Journal of Distributed Sensor Networks. 2015. DOI: 10.1155/2015/315714

[74] Hu Y, Leus G. Robust differential received signal strength-based localization. IEEE Transactions on Signal Processing. 2017;**65**(12):3261-3276. DOI: 10.1109/TSP.2017.2684741

[75] Caceres-Najarro LA, Song I, Kim K. Differential evolution with opposition and redirection for source localization using RSS measurements in wireless sensor networks. IEEE Transactions on Automation Science and Engineering October 2020;**17**(4):1736-1747. DOI: 10.1109/TASE.2020.2975287.

[76] Caceres Najarro LA, Song I, Kiseon K. Fundamental limitations and state-of-the-art solutions for target node localization in WSNs. 2021; TechRxiv. Preprint. DOI:10.36227/techrxiv.16698469.v1

[77] Sharp I, Yu K, Guo YJ. GDOP analysis for positioning system design. IEEE Transactions on Vehicular Technology. 2009;**58**(7):3371-3382. DOI: 10.1109/TVT.2009.2017270

[78] Fang X, Chen L. An optimal multi-channel trilateration localization algorithm by radio-multipath multi-objective evolution in RSS-ranging-based wireless sensor networks. Sensors.

2020;**20**(6):1798. DOI: doi.org/10.3390/s20061798

[79] Coluccia A, Fascista A. Hybrid TOA/RSS range-based localization with self-calibration in asynchronous wireless networks. Journal of Sensor and Actuator Networks. 2019;**8**(2):31. DOI: doi.org/10.3390/jsan8020031

[80] Le T et al. Hybrid TOA/AOA localization with 1D angle estimation in UAV-assisted WSN. In: 14th International Conference on Signal Processing and Communication Systems (ICSPCS); 14-16 December. 2020. Adelaide, SA, Australia: IEEE; 2021. pp. 1-6. DOI: 10.1109/ICSPCS50536.2020.9310043

[81] Zuo P et al. Directional source localization based on RSS-AOA combined measurements. China Communications. 2020;**17**(11):181-193. DOI: 10.23919/JCC.2020.11.015

[82] Bhat SJ, Santhosh KV. A method for fault tolerant localization of heterogeneous wireless sensor networks. IEEE Access. 2021;**9**:37054-37063. DOI: 10.1109/ACCESS.2021.3063160

[83] Bhat SJ, Santhosh KV. Priority based localization for anisotropic wireless sensor networks. In: IEEE International Conference on Distributed Computing, VLSI, Electrical Circuits and Robotics (DISCOVER); 30-31 October 2020. Udupi, India: IEEE; 2020. pp. 52-56. DOI: 10.1109/DISCOVER50404.2020.9278090

[84] Bhat SJ, Santhosh KV. Is localization of wireless sensor networks in irregular fields a challenge? Wireless Personal Communications. 2020;**114**: 2017-2042. DOI: 10.1007/s11277-020-07460-6

[85] Al-Turjman F. The road towards plant phenotyping via WSNs: An overview. Computers and Electronics in Agriculture. 2019;**161**:4-13. DOI: 10.1016/j.compag.2018.09.018

[86] Zhou G, He T, Krishnamurthy S, Stankovic JA. Models and solutions for radio irregularity in wireless sensor networks. ACM Transactions on Sensor Networks (TOSN). 2006;**2**(2): 221-262. DOI: 10.1145/1149283.1149287

[87] Phoemphon S, So-In C, Leelathakul N. A hybrid localization model using node segmentation and improved particle swarm optimization with obstacle-awareness for wireless sensor networks. Expert Systems with Applications. 2020;**143**:113044

[88] Pak JM, Ahn CK, Shi P, Shmaliy YS, Lim MT. Distributed hybrid particle/FIR filtering for mitigating NLoS effects in TOA-based localization using wireless sensor networks. IEEE Transactions on Industrial Electronics. 2017;**64**(6): 5182-5191

[89] Banihashemian SS, Adibnia F, Sarram MA. A new range-free and storage-efficient localization algorithm using neural networks in wireless sensor networks. Wireless Personal Communications. 2018;**98**(1):1547-1568. DOI: 10.1007/s11277-017-4934-4

[90] https://www.comsoc.org/publications/best-readings/reconfigurable-intelligent-surfaces

[91] Wymeersch H, He J, Denis B, Clemente A, Juntti M. Radio Localization and Mapping with Reconfigurable Intelligent Surfaces: Challenges, Opportunities, and Research Directions. IEEE Vehicular Technology Magazine. 2020;**15**(4):52-61. DOI: 10.1109/MVT.2020.3023682

[92] Hu S, Rusek F, Edfors O. Beyond ma ssive MIMO: The potential of positioning

with large intelligent surfaces. IEEE Transactions on Signal Processing. 2018; **66**(7):1761-1774

[93] He J, Wymeersch H, Kong L, Silvén O, Juntti M. Large intelligent surface for positioning in millimeter wave MIMO systems. In: IEEE 91st Vehicular Technology Conference (VTC2020-Spring); 25-28 May 2020. Antwerp, Belgium: IEEE June; 2020. pp. 1-5

Section 2

WSN and Internet of Things

Chapter 2

IoT-Based Decision Support System for Health Monitoring of Induction Motors

A. Pravin Renold and B. Venkatalakshmi

Abstract

An electrical motor is a common device that is used for a variety of electrical purposes. Because of their wide range of applications, motors that are both reliable and long-lasting are in high demand. Motors are prone to a variety of faults, including rotor bar breaking faults, short turn faults, bearing outrace faults, and so on. Unexpected faults or failures in these motors reduce workplace productivity. The time it takes to resolve the issues reduces the organization's profit. Bearing failures account for approximately 42% of all faults. Due to continuous operation, the shape of the majority of electrical motors with rolling bearings becomes disproportional. This causes the motor's elastic limit to be exceeded, as well as fractures, vibrations, and a rise in temperature. A good solution is to switch from scheduled maintenance to predictive maintenance, which is based on monitoring the motor's operating condition. This chapter proposes an Internet of Things (IoT)-based solution that continuously monitors and records the vibration from the induction motor. A decision support system analyzes the impact of vibration using log data and the Naïve Bayes classifier. The proposed decision support system detects the critical level of vibration and notifies the user of the motor's abnormal working condition.

Keywords: health management, induction motor, mitigation measures, decision support system, Naïve Bayes classifier

1. Introduction

In this third era of computing, the access of predictive information from remote locations becomes essential for optimal business solutions. This is applicable for every field of applications such as agriculture, manufacturing, healthcare, education, etc. The thumb rule for the achievement of the above information access is the development of an interface between recent technologies along with proven techniques of prediction and communication. This chapter presents one such effort of interfacing Internet of Things (IoT) technology along with conventional sensing and prediction techniques. The potential solution presented in the chapter can find its role in all possible fields, which uses induction motors.

The chapter framework is composed of four major sections. The following section is an introductory part highlighting the fundamentals of various induction motor faults and the basics of IoT technology. Section 2 of the chapter describes the problem statement along with the solution using the IoT architecture. The predictive maintenance designed as a decision support system using Naïve Bayes classifier algorithm has been elaborated in Section 3. Section 4 of the chapter highlights the chapter's outcomes and the performance metrics for the proposed IoT-based solution.

1.1 Fundamentals

The chapter deals with optimal solution of induction motors and smart monitoring and precaution effects of induction motors. The key fundamentals required for such architecture are the induction motor fundamentals and the Internet of Things fundamentals. So in this section, let us review the basics of them.

1.1.1 Construction of single-phase induction motor

The single-phase induction motor has a stationary part called stator and the rotating part called rotor. The stator made up of stampings carries the winding called stator winding. It is excited by a single-phase AC supply. The number of poles (P) for which stator winding wound decides the synchronous speed of the motor. The synchronous speed is given by Ns.

$$Ns = 120f/P \text{r.p.m.} \tag{1}$$

where f is the frequency. The induction motor always rotates at a speed that is slightly less than the synchronous speed. The rotor is a rotating part of induction motor. The rotor is connected to the mechanical load through the shaft. The rotor of the three-phase induction motors is further classified as:

i. Squirrel cage rotor

ii. Slip ring rotor or wound rotor or phase-wound rotor

Depending upon the type of rotor used, the three-phase induction motor are classified as:

i. Squirrel cage induction motor

ii. Slip ring induction motor or wound induction motor or phase-wound induction motor.

1.1.2 Working principle of induction motor and different types of faults

When AC supply is given to stator winding, it carries an alternating current, which produces the alternating flux. This flux links with the rotor conductors and due to mutual induction, the rotor experiences induced e.m.f. The rotor current produces another flux called as rotor flux, which is required for motoring action.

Like every machine, the electrical motors are also prone to various faults under different operating factors and lifetime. The commonly arising problems in electrical motors are [1] as follows:

i. Stator faults

ii. Broken rotor bar or cracked rotor end-rings

iii. Air-gap irregularities

iv. Shorted rotor field winding and

v. Bearing failures

Most of the electrical motors use rolling bearing, which is used for the smooth rotational movement of the rotor. A bearing consists of two rings, one inner and the other outer. A set of balls placed in raceways rotates inside these rings [2]. The bearings are affected by the stress caused by vibration, eccentricity, and bearing currents [3]. Around 40–50% faults in electrical motors are bearings-related [2]. Energy consumption, revolutions per minute, temperature, air gap eccentricity, vibration, and bearing health are some of the useful data claimed by various sensor manufacturers in relation to electric motors. This type of information can be useful in troubleshooting failed motors, inspecting the condition of operational motors, and determining when a motor requires a closer look or maintenance and to reduce the electric motor downtime.

The usage of IoT to collect data and analyze the sensed data helps to obtain knowledge from the raw data. Such information can be used to achieve predictive maintenance of electric motors. Some of the applications are turbines, paper mills, refrigeration, to name but a few.

1.1.3 Wireless sensor network (WSN)

Wireless sensor network (WSN) is one of the enabling technologies of Internet of Things. WSN is defined as a network of distributed sensor nodes which performs the task of sense, compute, and communicate. The WSN is an example of infrastructureless, short-range, personal area network. The communication standard for WSN is IEEE802.15.4. The WSN supports different types of topology such as star and mesh. The network comprises source nodes to monitor the environmental factors and forward the sensed data toward the sink node via multihop communication.

An approach for routing IPV6 packets over zigbee-based WSN is called the 6lowpan (IPV6 over low-power wireless personal area networks) [4]. The IP packets are compressed using the adaptation layer to make it suitable for the personal area network. With the help of border gateway, which acts as an interface between internet and the nodes in the sensor networks, the packets have been adapted suitable for Internet and wireless sensor networks.

1.1.4 Internet of Things (IoT)

Research trends in pervasive computing technologies have the principle of integrating the paradigm of many recent technological solutions for developing anytime, anywhere accessible devices and systems. This requires a major backbone technology as IoT. The master's and doctoral learning community, while exploring the design and applications of IoT, needs to establish new frameworks and integrate various ideas.

Internet of Things (IoT) leads in the world news today due to its wider potential of applications such as smart cities, smart homes, wearables, automobile industries, etc. Research in the field of IoT cannot be confined to a specific area [5]. It is enabled by handshaking of several domains of research such as sensors, networking, cloud computing, edge computing, big data, machine learning, and deep learning. As IoT is a technology outcome of multidisciplinary research, today's researchers are in need to develop Proof-of-Concept (POC) solutions on various aspects of IoT. There exists a significant tool for the design and development of IoT networks and solutions. For example, Contiki OS is a platform that has well-structured functions and modules supporting various design aspects of IoT networks. Usage of the features of communication stack of such tools provides extended IoT applications. The network layer of IoT can be enhanced with scheme of use of network coding at packet level to improve the throughput performance of IoT networks. Similarly, the security in the network layer of IoT can be enhanced by developing privacy homomorphism. Also such design can be enhanced with improvised Routing Protocol for Low-Power and Lossy Network (RPL) in the network layer for higher performances. Various applications can also be developed by generating new suitable functions of various algorithms as embedded solutions. **Figure 1** shows the architecture of Internet of Things where the hardware components are interconnected by the means of Internet. The following are the various benefits of IoT:

a. Communication between devices: IoT achieves the communication between devices, also famously known as Machine-to-Machine (M2M) communication. The devices are connected to a network, and hence, the control and transparency are available with greater efficiencies and quality.

b. Automation and control in working: As the devices are connected to a network, the devices can be controlled in a centralized manner with a widely used wireless technology called Wireless Fidelity (Wi-Fi).

c. Improved quality in monitoring of devices: IoT allows automating the tasks with less human intervention. The continuous monitoring leads to improved quality in decision-making, transparency, and quick decision-making during emergency situations.

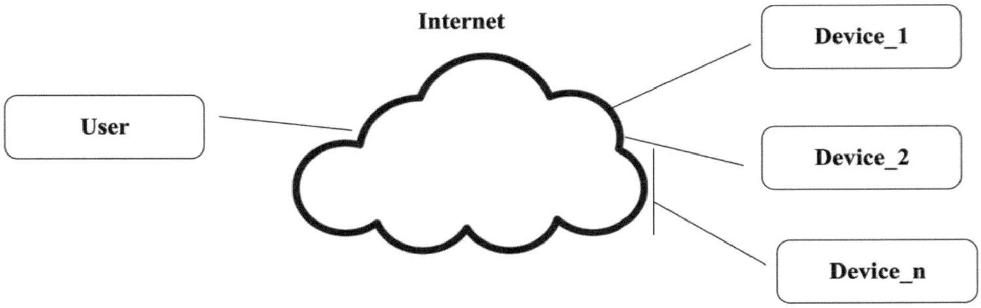

Figure 1.
Architecture of IoT.

2. Fault recovery analysis of induction motors

The induction motor finds application in most of the industries. Bearing fault in induction motor leads to more severity if not rectified in the initial stage [6]. The occurrence of bearing fault causes increased vibration and temperature of the motor. When the vibration goes beyond a certain level, it affects the air gap between stator and rotor and induces faulty frequency into the stator current. Many researchers have analyzed the faults in induction motors and proposed different strategies for monitoring and diagnosis. The following are few such existing solutions.

2.1 Online motor condition monitoring system for abnormality detection

An online motor condition monitoring system based on Cortex-M4 microcontroller with a graphic user interface is used for abnormality detection [7]. The system monitors the electrical and vibration signal for fault detection. The parameters monitored are voltage, current, and vibration. The captured signals are given to an infinite impulse file, and then fast Fourier transform is applied for spectral analysis to identify any abnormality in the captured signals pattern.

2.2 An analytical approach of parametric monitoring of induction motor using GSM

An embedded system based on ATMEGA-16 with Global System for Mobile Communication (GSM) has been used to protect the induction motor against overvoltage, overcurrent, and over-temperature [8]. The components such as timer, contactor, voltage, and current relays are used. The parameters used for finding the fault in the system are voltage, current, speed, and temperature. The parameters associated with the induction motor are collected for every periodical interval, the data are transmitted over GSM, and the messages are displayed in a Liquid Crystal Display (LCD) on the receiver end. Also the values are displayed on the mobile phone associated with the devices.

2.3 Acoustic based on fault diagnosis in induction motor

The work in [9] discusses an acoustic-based condition monitoring and fault diagnosis-based review to detect four different types of faults such as bearings, rotors, stators, and compounds. Various datasets are being analyzed using various machine learning algorithms. The type of fault determination in the induction motor is affected by environmental noise.

2.4 IoT-based vibration monitoring

The accelerometer sensor, which was mounted on the engine's rod axis and linked to a wireless RF device, was carried in different environments for different rotational speeds [10]. Furthermore, various types of vibration signals with varying amplitudes and frequencies are injected directly on the engine's axis to test and prove device reliability. Allan's variance technique allows for the successful detection, definition, and localization of vibration signatures.

2.5 Health monitoring using IoT and machine learning

A real-time machine health monitoring system that can analyze the supply balancing condition on a three-phase system by combining machine learning and IoT technology is proposed in [11]. This system is built with current transformer to capture and send electrical data from the load to the server. The server processes data by artificial neural network to train the data and for load classification.

2.6 Case study on fault diagnosis in induction motor

Incorporation of machine learning algorithms to aid or to take decision on its own on different types of faults in induction motor. In [12], different artificial intelligence algorithms and its suitability for fault identification have been discussed.

- Neural network, after the training phase, is used to classify the incoming data. The value that lies outside the range is named as a potential motor fault. To avoid false fault diagnosis, the alarm is raised when fault value ranges are observed persistently. It is suitable to diagnose bearing and unbalanced rotor faults of induction motors.

- Fuzzy-logic-based systems have been used to classify broken-bar-related. A set of nine rules are used to determine the two sideband components. The broken bars are identified based on the sideband components.

- A spectral kurtosis and envelope spectrum to identify different types of faults in rolling element bearings [13]. The dataset [14] contains an acceleration signal "gs," sampling rate "sr," shaft speed "rate," load weight "load," and four critical frequencies representing different fault locations such as ballpass frequency outer race, ballpass frequency inner race, fundamental train frequency, and ball spin frequency.

2.7 Inference

Fault detection and diagnosis are an aiding tool for the accurate determination of different types of fault in induction motor. The efficiency of the fault diagnosis system depends on the system or algorithm accuracy. Fault prediction is capable of predicting early possible development of fault in the induction motor. It leads to reduced maintenance cost and less shut down time of the equipment. Fault prediction system in association with Internet of Things could be an effective method to continuously track the status of the equipment and allocate periodical prior schedule of the equipment from a remote location.

3. Fault prediction model

The prediction model is designed with Naïve Bayes algorithm. Naïve Bayes classifier predicts that the presence (or absence) of a particular feature of a class is unrelated to the presence (or absence) of any other feature [15]. This classifier is very simple, efficient, and is having a good performance. Sometimes it often outperforms more sophisticated classifiers even when the assumption of independent predictors is far. This advantage is especially pronounced when the number of predictors is large.

3.1 Overall system

The system is designed with the objective of identifying the fault condition in the induction motor and informing the status of the induction motor to the lab in-charge in a remote fashion. The advantage of this approach is that the induction motors located at different premises are monitored, and periodical maintenance is allocated in a centralized manner as shown in **Figure 2**. The induction motor equipped with accelerometer to monitor the vibration data and the data are sent to the gateway node in a wireless fashion. The gateway node forwards the data to the control room for every periodical interval. The decision-making software runs the Naïve Bayes algorithm on the received data. The algorithm predicts the possible occurrence of fault. The alarm will be given to the lab in charge for further maintenance if any.

Figure 3 shows the proposed prediction technique. The induction motor is connected with accelerometer. The accelerometer data are forwarded to the gateway and to the control room server. The server runs the Naïve Bayes algorithm for the purpose of occurrences of fault. The acceleration data are processed by the application created by python programming language. The Naïve Bayes algorithm implemented in python language has been trained with the accelerometer data. After testing with different samples of data with both normally working machine and with faulty induction motor, the predictor module predicts the status of the machine from incoming real data after during the real-time running of the motor. The outcome of the predictor is divided into two classes, namely normal and faulty class. If faulty class is predicted, then an alert is given to the lab in-charges. Hence, preventive maintenance is carried out to avoid long time stoppage of the motor, thus improving the production time of the induction motors.

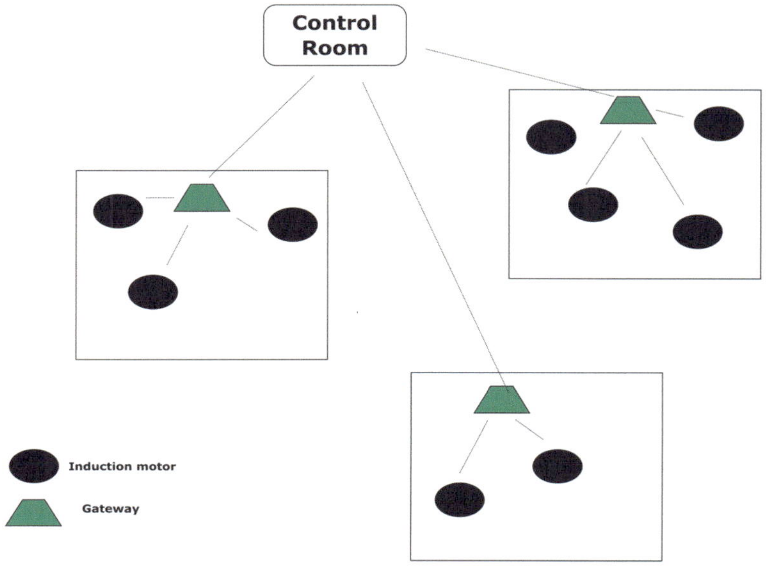

Figure 2.
Overall system setup.

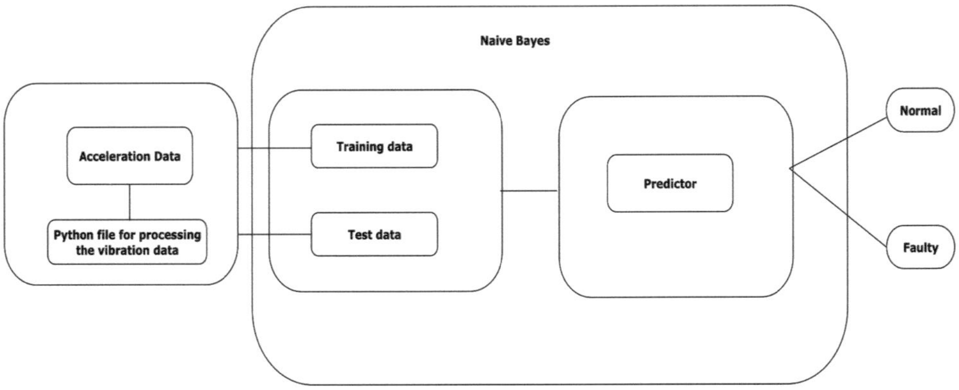

Figure 3.
Proposed decision-making technique.

3.2 Modules of prediction model

Figure 4 shows the module of the proposed technique. The detailed explanation of the modules is given in this section.

Figure 5 shows the experimental setup for measuring the vibration of the induction motor. An embedded base board with Wi-Fi support known as Intel Edison is connected with the accelerometer using the Arduino's Uno board, which acts as data acquisition unit. The vibration data from the induction motor are collected for every periodical interval. The sensed data are forwarded to the lab in-charge, to know the status of the machine. Also, the data are processed with Naïve Bayes algorithm for the determination of bearing fault.

Based on the outcome of the decision-making module, which runs the Naïve Bayes algorithm, the lab in-charge decides whether to allow the motor to run or to stop or halt the motor for a specific period of time. The hardware and software used for the implementation are tabulated in **Table 1**.

The embedded device we have used in the work is Sparkfun Intel Edison board [16]. It is a lightweight board designed to support Internet of Things, and the base board consists of 70 pins. It has inbuilt Wi-Fi and bluetooth, it supports Yocto Linux operating system. The operating voltage of the board is 3.3–4.5 V. The application development could be done by Python, Jjava, Node.js, C, C++. The application development on the Intel Edison could be done by performing the connection via putty software.

The Intel Edison board is connected with Arduino Uno using Inter Integrated Circuit (I2C) protocol as shown in **Figure 6**. The Intel Edison board is the master, and the Arduino Uno is the slave. The reason for adapting Arduino Uno is its easy interfacing support with sensors. I2C is a synchronous serial protocol, the clock signal is controlled by the master as shown in **Figure 6**. The serial data (SDA) line is used for the master and slave to send data. The serial clock (SCL) line carries the serial clock. In I2C, the messages are broken into frames. Each message has an address frame that contains the binary address of the slave, and one or more data frames that contain the data being transmitted. The message also includes start and stop conditions, read/write bits, and ACK/NACK bits between each data frame.

Figure 7 shows the interfacing of ADXL335 with Arduino. ADXL335 has physical vibrations as input and the three-axes analog output is taken via the X, Y, and Z pins from the accelerometer. The Arduino UNO has A1, A2, and A3 as analog input/output

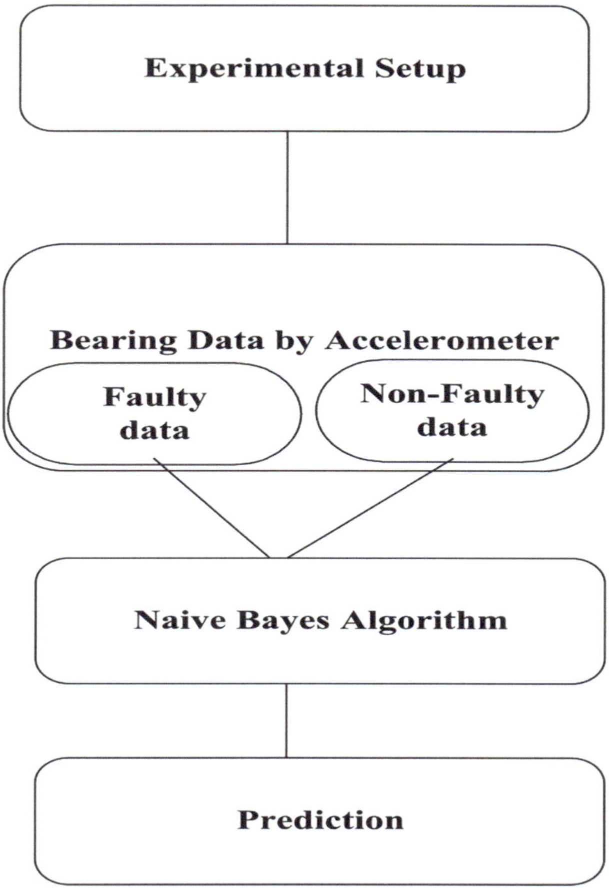

Figure 4.
Modules of proposed technique.

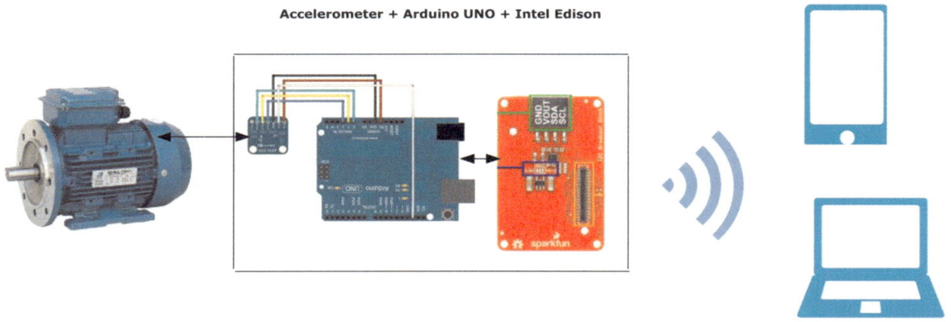

Figure 5.
Experimental setup.

pins. Accelerometer output is given as input to the Arduino by connecting X, Y, and Z to A1, A2, and A3, respectively. The Aref pin of the Arduino is shorted with the 3.3 V pin of the Arduino itself. Accelerometer receives supply from the Arduino by connecting the Vcc and GND pins of the both with each other.

Hardware	Software
Single-phase induction motor	Python (Application and Device Driver)
ADXL 335 3-axis Accelerometer	Arduino Code (Dialect of C/C++)
Arduino UNO	HTML
Sparkfun Intel Edison	Naïve Bayes algorithm

Table 1.
Components for vibration monitoring.

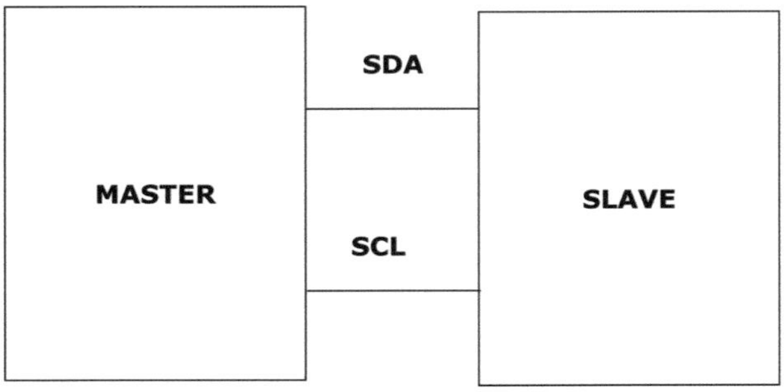

Figure 6.
I2C communication between master and slave.

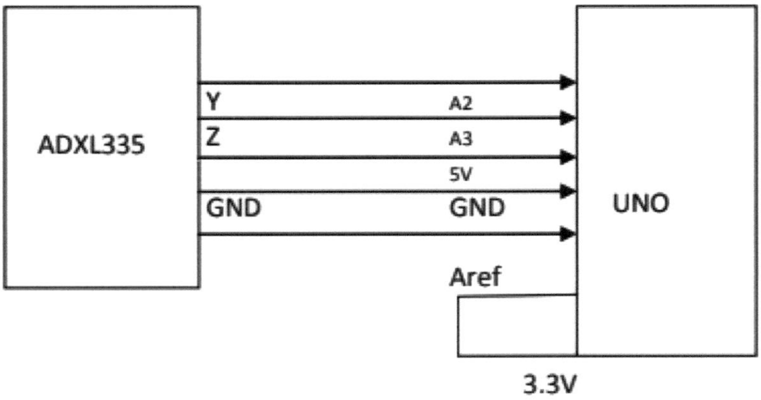

Figure 7.
Interfacing accelerometer sensor with Arduino Uno.

Figure 8 shows the interfacing of Arduino Uno with Intel Edison. The accelerometer values are taken to the Intel Edison board via I2C protocol. A4 and A5 pins of the Uno are the analog input/output pins. I2C board block of the Intel Edison block has SCL and SDA pins for serial communication. A4 and A5 of the Arduino are connected to SDA and SCL, respectively. The GND-GND connection of both the board needs to be ensured.

- Prediction analysis and solution: Naïve Bayes classifier

Predictive maintenance is a method to predict the occurrence of fault in the system in advance. The advantage of such approach is the time and cost involved in

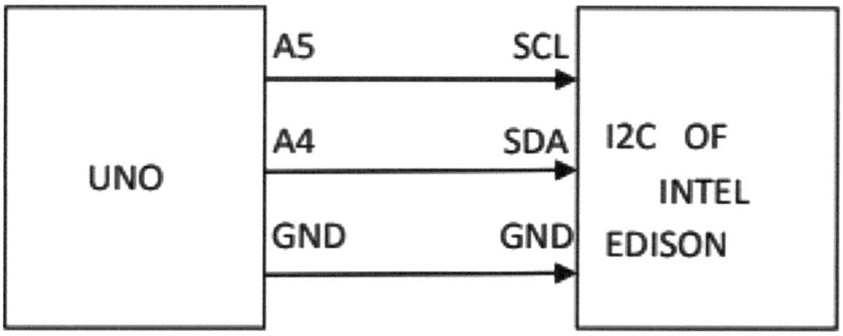

Figure 8.
Interfacing Arduino Uno with Intel Edison.

overcoming the fault is less compared with the traditional approaches such as run to failure and preventive maintenance. There are different prediction algorithms used to determine the fault or anomaly in advance. In [17], comparison of supervised machine learning algorithms for IoT data has been performed. Algorithms such as Naïve Bayes, decision tree, random forest, k-nearest neighbor, and logistic regression are being used for different datasets. The summary is as follows:

- The Naïve Bayes is suitable for moderate size of dataset.
- As the number of features increases, the time taken for the result is increased in all algorithms.
- For better performance, the number of classes needs to be kept minimal in Naïve Bayes, whereas in decision tree, random forest, and k-nearest neighbor, the number of classes do not have impact on the performance.
- Naïve Bayes performs well even with missing values in the dataset.

The advantages of Naïve Bayes algorithm are as follows:

i. Simple and fast

ii. Requires less training data

iii. Suitable for continuous and discrete data

iv. Suitable to make probabilistic predictions

The Naïve Bayes classifier is based on Bayes theorem and is used to find the probability of each instance belongs to a specific class. It operates on dataset X with attribute values $\{x_1, x_2, \ldots, x_n\}$ and determines the target function Y(X) from a predefined set $S = \{s_1, s_2, .., s_n\}$

$$P(S|X) = \frac{P(X|S)P(S)}{p(X)} \tag{2}$$

$P(S)$ is the prior probability about the class S. $P(X|S)$ is the posterior probability of X when the class S has given. The probability of obtaining result X without knowing S has occurred is $p(X)$, and $P(S|X)$ is the posteriori probability of X.

4. Simulation results

To evaluate the performance of the prediction system, we use python to develop the application for Internet of things, device driver code, and also the Naïve Bayes algorithm. To conduct the simulation, we used the dataset with and without fault. The dataset has been prepared by running the induction motor with and without bearing faults for different time intervals. The acceleration data were collected for an interval of 20 s. Some of the collected data are used for training to determine the threshold using Naïve Bayes model.

To classify the class from the new data, we have chosen the faulty value (1 or 0)

$$classifier(x_1, x_2, \dots, x_n) = \underset{s \in \{0,1\}}{\operatorname{argmax}} P(x_1, x_2, \dots, x_n|S) \tag{3}$$

The relative frequency counting according to multinomial Naïve Bayes is given as

$$P(X|S) = \frac{N_{xi,si} + a}{N_{si} + ad} \tag{4}$$

where $N_{xi,si}$ is the number of times the features x_i appears in class s_i; a is the parameter for additive smoothing; d is the dimensionality of the feature vector.

4.1 Data collection setup

The vibration data are collected using accelerometers, the vibration at different positions of the motor housing is studied. The position near the shaft is chosen for measuring the vibration. The vibration values are collected for different scenarios such as no load and moderate load. The accelerometer data are collected and forwarded to the server using the Intel Edison embedded board. Ten datasets are used in this study, which includes normal and faulty scenario. The faults may be any of the types such as inner race, ball defect, train defect, and outer race. All the 10 datasets are separated to form the new dataset in a fixed length of 800 after performing data cleanup. The label of the samples $Y = \{0, 1\}$ and the definition of the element are shown in **Table 2**.

4.2 Results

Table 3 shows the overall accuracy of the prediction algorithm to predict the possibility of fault in induction motor. Under no load scenario, the fault prediction is

Label	Status
0	Normal
1	Possibility of fault

Table 2.
Labels for classes.

	No load	Load
Accuracy	100%	93.2%

Table 3.
Accuracy.

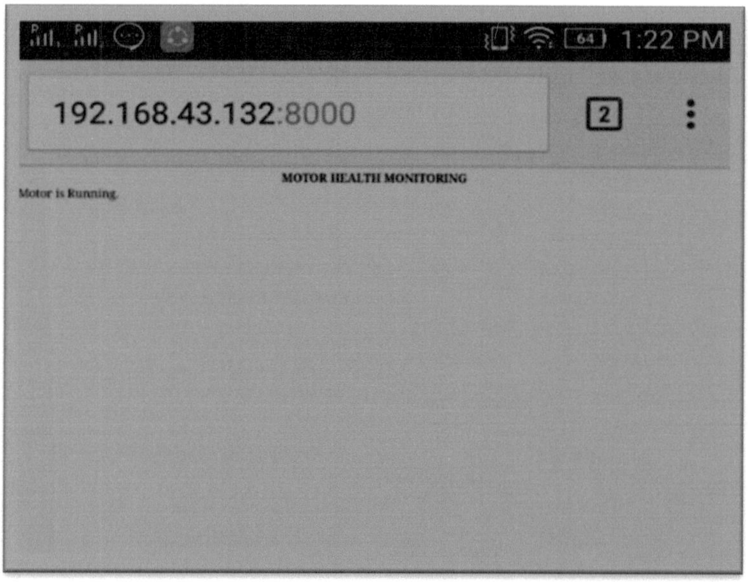

Figure 9.
Status of motor observed using an URL.

accurate, whereas during running condition, the accuracy is reduced, it could have been avoided by increasing the training samples and to have dataset collection for different variants of load, as we focus on the presence of fault or not, we are done with a moderate level of data for training and prediction.

Figure 9 shows the status of the motor in a web page using a dedicated address. The web page is updated with the status of motor running; if any fault is expected, an update is given to the server and the web page of the lab in-charge is automatically updated.

5. Conclusions

This chapter presented the design and validation of an IoT system for monitoring the health of induction motors in a laboratory setting. The system incorporates the use of a sensor in the IoT board that wirelessly transmits the sensed data, and the status of the induction motor after evaluation with the machine learning algorithm known as Naïve Bayes is made available on the web page. The accuracy of the developed system was evaluated under no load and load conditions.

Future work in this domain is to extend the work with more nodes and to consider many induction motor parameters to predict the fault in the induction motors by using a bag of learning model.

Conflict of interest

The authors declare that they have no conflict of interest.

Author details

A. Pravin Renold[1*] and B. Venkatalakshmi[2*]

1 Mepco Schlenk Engineering College, Sivakasi, India

2 L&T EduTech, Chennai, India

*Address all correspondence to: pravinrenold@mepcoeng.ac.in
and venkatalakshmi.71@gmail.com

References

[1] Vas P. Parameter Estimation, Condition Monitoring, and Diagnosis of Electrical Machines. New York: Oxford University Press; 1993

[2] Kliman GB, Stein J. Induction motor fault detection via passive current monitoring. In: International Conference in Electrical Machines. August 13, 1990; Cambridge, MA. 1990. pp. 13-17

[3] Chen S, Lipo TA. Bearing currents and shaft voltages of an induction motor under hard-and soft-switching inverter excitation. IEEE Transactions on Industry Applications. 1998;**34**(5): 1042-1048

[4] Rana B, Singh Y, Singh PK. A systematic survey on internet of things: Energy efficiency and interoperability perspective. Transactions on Emerging Telecommunications Technologies. 2021;**32**(8):e4166

[5] Armentano R, Bhadoria RS, Chatterjee P, Deka GC, editors. The Internet of Things: Foundation for Smart Cities, EHealth, and Ubiquitous Computing. Boca Raton, Florida: CRC Press; 2017

[6] Kompella KD, Rao MV, Rao RS. Bearing fault detection in a 3 phase induction motor using stator current frequency spectral subtraction with various wavelet decomposition techniques. Ain Shams Engineering Journal. 2018;**9**(4):2427-2439

[7] Chang HC, Jheng YM, Kuo CC, Huang LB. On-line motor condition monitoring system for abnormality detection. Computers & Electrical Engineering. 2016;**51**:255-269

[8] Lande S, Jaiswal P, Rajgure P. An analytical approach of parametric monitoring of induction motor using GSM. IOSR Journal of Electronics and Communication Engineering. 2012;**1**(3): 01-07

[9] AlShorman O, Alkahatni F, Masadeh M, Irfan M, Glowacz A, Althobiani F, et al. Sounds and acoustic emission-based early fault diagnosis of induction motor: A review study. Advances in Mechanical Engineering. 2021;**13**(2):1687814021996915

[10] Hayouni M, Bousselmi Z, Vuong TH, Choubani F, David J. Wireless IoT approach for testing in situ motor's axis vibration monitoring. In: 2021 International Wireless Communications and Mobile Computing (IWCMC); 2021 June; IEEE. 2021. pp. 92-97

[11] Wong TK, Mun HK, Phang SK, Lum KL, Tan WQ. Real-time machine health monitoring system using machine learning with IoT technology. In: MATEC Web of Conferences, Vol. 335. EDP Sciences. 2021. p. 02005

[12] Nandi S, Toliyat HA, Li X. Condition monitoring and fault diagnosis of electrical motors—A review. IEEE Transactions on Energy Conversion. 2005;**20**(4):719-729

[13] Matlab [Internet]. 2020. Available from: https://in.mathworks.com/help/predmaint/examples/Rolling-Element-Bearing-Fault-Diagnosis.html

[14] Bechhoefer E. Condition based maintenance fault database for testing diagnostics and prognostic algorithms. 2013. Available from: https://mfpt.org/fault-data-sets/ [Online accessed Oct-2021]

[15] Soni J, Ansari U, Sharma D, Soni S. Predictive data mining for medical

diagnosis: An overview of heart disease prediction. International Journal of Computer Applications. 2011;**17**(8): 43-48

[16] Sparkfun Intel Edison [Internet]. 2020. Available from: https://learn.sparkfun.com/tutorials/general-guide-to-sparkfun-blocks-for-intel-edison/all

[17] Khadse V, Mahalle PN, Biraris SV. An empirical comparison of supervised machine learning algorithms for internet of things data. In: 2018 Fourth International Conference on Computing Communication Control and Automation (ICCUBEA); August; IEEE. 2018. pp. 1-6

Chapter 3

Network Slicing for Industrial IoT and Industrial Wireless Sensor Network: Deep Federated Learning Approach and Its Implementation Challenges

Seifeddine Messaoud, Soulef Bouaafia, Abbas Bradai, Mohamed Ali Hajjaji, Abdellatif Mtibaa and Mohamed Atri

Abstract

5G networks are envisioned to support heterogeneous Industrial IoT (IIoT) and Industrial Wireless Sensor Network (IWSN) applications with a multitude Quality of Service (QoS) requirements. Network slicing is being recognized as a beacon technology that enables multi-service IIoT networks. Motivated by the growing computational capacity of the IIoT and the challenges of meeting QoS, federated reinforcement learning (RL) has become a propitious technique that gives out data collection and computation tasks to distributed network agents. This chapter discuss the new federated learning paradigm and then proposes a Deep Federated RL (DFRL) scheme to provide a federated network resource management for future IIoT networks. Toward this goal, the DFRL learns from Multi-Agent local models and provides them the ability to find optimal action decisions on LoRa parameters that satisfy QoS to IIoT virtual slice. Simulation results prove the effectiveness of the proposed framework compared to the early tools.

Keywords: federated learning, industrial IoT, network slicing, QoS

1. Introduction

In the last decade, industrial manufacturing such as healthcare, smart grids based on Cyber-Physical Internet of Thing Systems (CPIoTS) has been widespread [1]. In this context, IIoT network, which is characterized by the unified network physical layer, the QoS constraints, the autonomous connection requirements, is considered one of the key issues. The rapid increase in data amounts with diverse QoS requirements [2] brings several challenges in order to meet the complex requirements, as well as resource and QoS requirements with high data rate and low latency. In fact, the advanced 5G technology has a significant potential to provide IIoT QoS satisfactory

[3]. With their architectural approaches which are founded on a unified physical layer, addressing the diverging performance requirements in terms scalability and availability still a hot challenges topic. Today's drastic digital transformations empowered by emerging technology like Edge Computing, Software Defined Networking (SDN), Network Function Virtualization (NFV), and LoRaWAN can bring smart services for network candidates [4]. Network slicing (NS) is the key solution that provides smart service's connectivity with diverse QoS requirements. Using deep RL at each LoRa agent in the environment. Each agent considered to be a Deep Q-learning (DQL) brain interacts with the environment to find the best action on their parameters that brings the best reward. In addition, it introduces the FL approach to provide better RL based action on each agent, to maximize QoS, and hence throughput revenue. However, NS provides the network availability as a service following the slice instances exploiting NFV and SDN [5]. In this context, a Mini Batch GD and GMM framework is proposed in [6] to provides radio resources for the virtual slice member. In addition, a LoRa network slicing technique based on Maximum Likelihood Estimation proposed, in [7], to allocate network resources in inter and intra mode. Meanwhile, recently supervised learning approach-based resource allocation is also proposed to manage network resources, but due to the training data unavailability or the high computational training process, are not appropriate for large-scale network and cannot satisfy dynamic slices requirements.

RL technique can improve efficient resource management by interacting with the environment, in which Q-learning is the widely used. The RL agent learns the association between taken action and the received feedback in terms of reward. It follows a policy, which is updated according to the maximized revenue via several action series. Therefore, high-quality policies building in a centralized network architecture faces a major challenge, especially when the space of state features is restricted. To deal with these issues, Federated Learning (FL) has been suggested as a decentralized tool for machine learning, which is designed to be a global learning system. In this context, the aim of this work is to propose a deep federated reinforcement learning (DFRL), to equip the slice member with the required channel resources, by tuning LoRa TP and SF parameters [8].

The leftover of this chapter is organized in six sections. Section 2 presents the related work of this chapter. In Section 3, we give a brief overview on the Industrial IoT, the federated learning, and the network slicing. We highlight, in Section 5, the proposed slicing architecture and the system model. After that, in Section 6, the relation between wireless sensor network (WSN) and the IIoT is well highlighted. Next, the proposed slicing resource reservation-based DFRL framework is presented in Section 6. Section 7 evaluates the simulation results. Finally, Section 8 concludes the chapter.

2. Related works

Recently, several articles have investigated the many challenges typical of the network slicing approach. In particular, in [9] the authors propose an online auction algorithm to realize a resource allocation framework, capable of guaranteeing the diversity of services to users and high levels of social welfare. Differently, the work in [10] deals with a new resource allocation framework to automatically and automatically size the capacity and size of network slices. In this chapter, resource partitioning is done based on both available network bandwidth and LoRa configurations parameters resulting in an optimal trade-off between traffic and network aspects. The authors of [11] focus on the design and implementation of a dynamic slicing sharing system to ensure minimum user throughput requirements. Therefore, it addresses three sub-issues: admission control, resource allocation, and user abandonment issues. The contextualization of the placement of VNFs to the network slicing problem is presented in [12], where, in particular, the topological information of the network is exploited to provide an appropriate deployment of functions, with regard to different service classes. Moreover, the placement problem of VNFs has also been addressed in [13], which considers the function decomposition and the sub-functions sharing, a profitable heuristic algorithm is proposed based on Linear integer programming formulation of the VNF placement problem. Further, in the work [14], a formulation of mixed integer linear program is exploited to process the number identification of VNFs to use, which aims to meet specific service requirements. To solve the placing VNFs problem in a federated cloud, the coalition formation game is proposed in [14]. Alternatively, a Pareto analysis of the VNF placement problem is the subject of [15]. In [16], a network partitioning policy is developed to take into account the social well-being and the supplier profit of the network. Finally, special cases for the VNF placement problem are discussed and analyzed in [17].

The problem of bandwidth slicing in software-defined networks is studied in [18], where price spikes are exploited to indicate the presence of traffic spikes and network congestion. This work provides a time-based price analysis combined with a Stackelberg game, in which the gain of SP Internet is the gain of income. In a different way, the work cited in [19] studies the correlation between the network slices size and the resource pricing strategy. In addition, an algorithm to vary the prices is proposed by the authors in [19], with the aim of maximizing both the customer profit and the SP. Although FL has not been used in the field of network slicing research, FL has recently reached attention and several papers have presented its use, the methods cited in [20, 21] being prime examples of such a branch of literature. In [20], a new aggregation data scheme for wireless computation is proposed. The strategy exploits the signal overlay property of the wireless channels. Differently, articles [20, 22] focus on maximizing the number of devices involved in the aggregation process, also taking into account the minimization of aggregation error. Thus, it contextualize the FL in an MEC system and apply the distributed gradient descent method to identify the best compromise between local updates and global aggregations, aiming to minimize the loss function, in taking into account certain resource constraints. Likewise, the article in [21] analyzes the MEC environment and presents the application of hybrid filtering on stacked encoders to predict fluctuation in file popularity in the content caching problem. Moreover, the article cited [23] modifies the proposed, federated averaging algorithm with the stochastic gradient descent algorithm, to train the data in a distributed way, thus reducing communication costs. The multitasking learning problem is studied in the work cited in [24], authors proposed a new Mocha contextual optimization approach that used in combination with the FL system. The work cited in [25] analyzed the End-to-end delay in a blockchain framework, in which an FL blockchain structure is developed to perform a distributed consensus strategy. In order to improve the transmission and computational costs in a hybrid IoT-MEC network, authors in [26] proposed to use the FL powered by the multiple deep reinforcement learning agents. In addition, ultra-dense scenarios are also considered in [27], where an approach based on the technique of deep learning of short-term long-term memory is applied to forecast local network traffic in order to avoid congestion.

This chapter aims to address the problem of network slicing using deep federated RL at each LoRa agent in the environment. Each agent considered to be a Deep Q-learning (DQL) brain interacts with the environment to find the best action on their parameters that brings the best reward. In addition, it introduces the FL approach to provide better RL-based action on each agent, to maximize QoS, and hence throughput revenue.

3. Overview on industrial IoT, federated learning and network slicing

The development and evolution of modern information and communication technologies lead us to the Fourth Industrial Revolution, in which the Industrial Internet of Things (IIoT) is supposed to be one of the key aspects to realize the industry 4.0. With an unprecedented increase in the number of Internet of Things (IoT) devices and emerging applications, a large amount of traffic is created every day. Such an increase represents a heavy load on the Internet network and also requires significant investments for the upgrade of the infrastructure. However, with the development of big data analysis and artificial intelligence (AI) techniques such as deep learning (DL) and machine learning (ML), the data collected can be effectively exploited for many purposes. From a communication point of view, the last few years have seen the emergence of AI applications in various fields. For example, ML is used to study efficient antenna selection in multi-antenna wireless systems [28], DL is used to handle the computational offload problem in IoT systems with edge computing [29], and Deep reinforcement learning (DRL) is used to optimize resource allocation issues at the edge of the network, such as traffic classification, edge caching, network security, and data offload [30]. However, conventional AI models generally require central processing of the data collected from all users on the network, i.e. users have to upload their own data to a central server to train the learning model. However, a key concern with central learning is data privacy, i.e., some users want to keep track of their local data and do not want to transmit their local data to the central server. Training the learning model centrally requires a central cloud with extremely powerful compute and storage capabilities. Meanwhile, recent advancements in computer hardware and the proliferation of smart devices in our daily lives have shown that every IoT device can be equipped with reasonable levels of compute and storage, which is closely comparable to a desktop computer there was. is 10 years old [31]. Therefore, the standard ML model is not easily applicable to large scale IoT networks and cannot exploit the availability of distributed computing. This requires a new learning model that leaves training data distributed across individual IoT devices instead of being centralized.

Motivated by this problem, Google invented the concept of federated learning (FL) for on-device learning and data privacy preservation [32]. Using the FL approach, each IoT device can train its model based on locally collected data. Local data from IoT devices does not need to be sent to the centralized cloud. The centralized cloud only needs to collect the updated local training model from individual users. Due to its characteristics, FL has been adopted in many applications, for example FL for improving Google keyboard suggestions [33], FL for healthcare [34], and FL for smart city detection [35]. To illustrate the concept of FL, an overview of FL in IoT systems is shown in **Figure 1**. In general, each IoT device has its own set of data and the aggregation server can either be located at the edge of the network or in a virtual cloud in the remote cloud computing system [36]. Each FL model has its own advantages and disadvantages, depending on various factors. For example, FL with the server at the edge of the network is suitable for applications requiring low latency, location awareness and contextual information on the network while cloud-based FL is suitable for applications with IoT devices massive over multiple regions and computing power requirements/storage capacities.

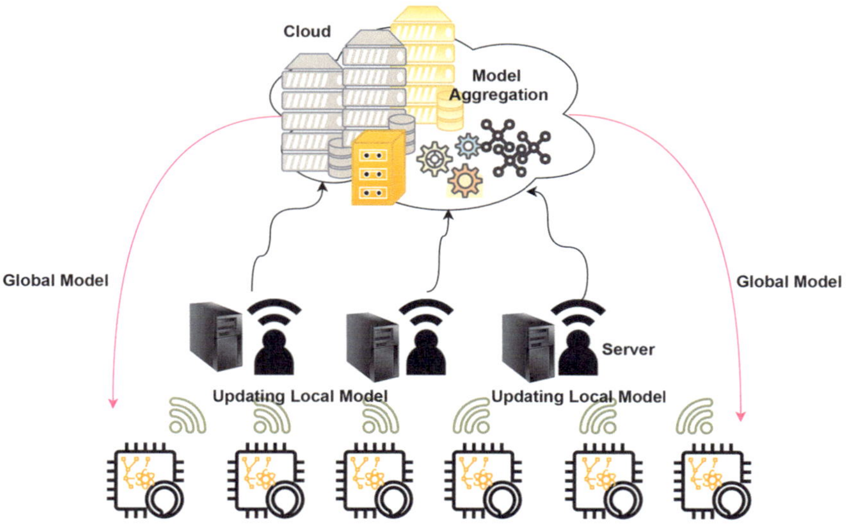

Figure 1.
An overview of FL in IoT systems.

Recently, the integration of DL models with IoT and edge devices has become more popular, which provides real-time analytics with limited resources. Thus, Federated DL (FDL) allows Industry 4.0 companies to integrate DL into IoT devices and provides a secure framework using FL, as shown in **Figure 2**. DL is computationally expensive, which requires resources and an expensive framework. Thus, the decentralization of DL models is a multidimensional problem that requires a framework of new technologies to integrate DL with advanced computing and the IIoT. The main goal of FDL is to provide the IIoT with advanced capabilities using optimized DL that would turn Industry 4.0 factories into smart factories. Some of the parameters required to create the FDL model in IIoT are the FDL model, FDL networking, FDL security, and FDL optimization.

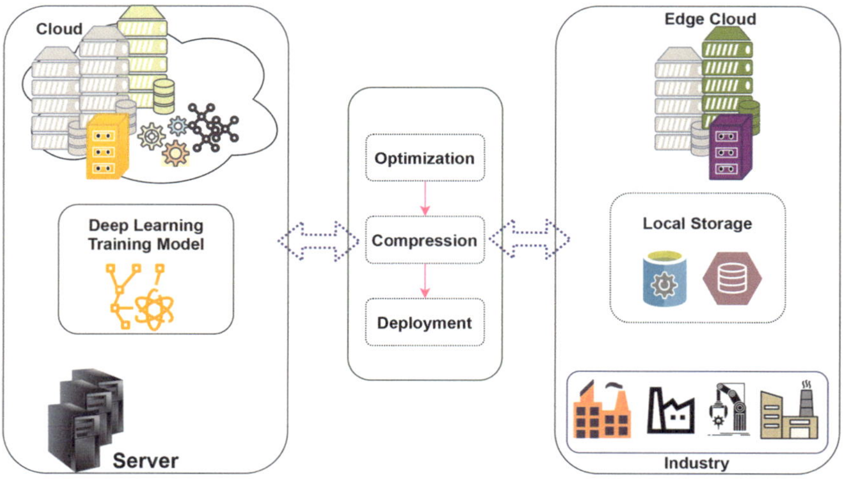

Figure 2.
Federated DL in IIoT.

An FDL model can be implemented on both client side and server side. On the client side, private networks are defined, the DL model of which is tuned and optimized from the general model present in the cloud. Optimized and fitted models are then deployed on the client side, where the model is trained with data generated locally from the end device. Finally, the final device contains a highly quantized and compressed FDL model. On the server side, the model in the cloud is continuously updated by differentially integrating the gradients of each private network. Each local DL network in turn is responsible for continuously uploading and uploading the currently updated gradients to the cloud model. Thus, a distributed selective stochastic gradient descent approach is presented in [37] which can be applied in the cloud model to frequently update the local private model. The first decentralized model called"Model chain" uses blockchain technology [38, 39] to allow the preservation of confidentiality in the transfer of data. In addition to this, asynchronous stochastic gradient descent can also be used when a single model can be trained in parallel among all devices, aggregated and processed.

Regarding FDL communication and networking, is that the main benefit of using FDL is to run DL models in IoT devices and involve the model in the decision making process. This type of decentralized DL process improves the robustness, operational efficiency and reliability of IoT devices. FDL provides two types of communication, namely intra-communication channel and inter-communication channel. Train transmits data between all levels of the framework. FDL communicates between the IoT and the cloud tier where the cloud-optimized model is deployed on the end device. However, security and confidentiality must be maintained in the FDL during communication. In inter communication channels, the components of each layer communicate with each other in three different ways, such as cloud, edge, and end device. The main objective of FDL is to minimize intra-communication and to maximize inter-communication, which would greatly reduce the cost of communication. By the way, to maintain privacy and security, FDL builds DL models that do not expose information about the data to the cloud. Security issue on the server side includes sharing of DL models on the cloud that leads to confidentiality and security risk. Security issue on the client side is done by encrypting the data during the training process before sending it to the cloud server. Some mechanisms and homomorphic encryption technique controls the amount of data to be shared on the cloud. Since peripherals have limited memory and computational requirements, DL models must be optimized so that they can be deployed to IoT or peripherals efficiently. In terms of hardware optimization, the GPU provides low-power computation that reduces computation time. The FPGA and Google's Tensor Processing Unit [40] are other DL devices that enhance DL network processing. In terms of memory optimization, algorithms such as shared memory allocation algorithms for DL models can be used. Dynamic scheduling [41] is one of the main processes used to improve performance on a cloud server.

4. Wireless sensor networks and its relation to industrial internet of things

4.1 Relation between IIoT and IWSN

At the heart of the IIoT are the WSNs, that include of multi-functional nodes, low-cost, along with sensing, have both communication and processing capabilities. In order to communicate wirelessly over short distances, these little, inexpensive sensor nodes have built-in transceivers and processors. They are densely exploited in an area of interest to collect sensory data, by coordinating and collaboratively exchanging information by training ad hoc wireless networks. Due to the small size and the batteries use, Sensor nodes are limited in processing, communication, and power. A unique feature of WSNs is their network processing attribute, whereby sensor nodes do not send raw sense data directly to the gateway but merge it locally to make it more consistent and save significant communication costs. Their application field is multiple and they are now ubiquitous components of intelligent environments, due to their unique attributes. Their various area of application covers home, surveillance, military, smart city, patient health monitoring, automation, etc. WSNs are used in telehealth applications in patient healthcare monitoring scenarios. As example, to monitor patients with chronic diseases and regularly check their various parameters such as heart rate, blood sugar and send this information wirelessly to a doctor remotely for further diagnosis. In order to help the elderly and disabled in their daily tasks, the WSNs are also used.

Indeed, they have seen major deployments in a diversity of applications, including agriculture, industrial process automation and control, transportation, and supply chain management over the past decades. Due to their ubiquitous presence and considering the potential benefits of these networks, such as simple deployment, cheap installation cost, no cabling cost, less complexity and mobility, they are increasingly used. in IIoT applications, which gave rise to IWSNs. WSNs can be used in an IIoT environment such as automation and control, process monitoring, and safety and emergency applications. In automation and process control applications, several tasks may require active nodes named actuators, which have the capability to act autonomously on the physical environment based on the detected measurements. For example, in the automation and control of feedback-based chemical processes, sensors measure temperature; if the temperature crosses a certain threshold value, they inform the actuators to reduce the temperature to a desired value so that the process remains in a stable state. Such applications place strict constraints on low latency and reliability because the sensor measurements must reach the actuator in a timely and reliable manner in order for the valve control action to be performed on time [42].

Today's sensor nodes have more processing power, longer battery life and memory, due to recent technological improvements compared to the first resource-constrained sensor nodes. This allowed them to be used in IIoT applications and resulted in IWSN. IWSN makes processes independent and autonomous, especially in difficult areas, to get actuation and control information, sensory. Sensor nodes in the WSN field detect process variables (e.g. temperature, pressure, etc.) and pass them to the well or gateway. The sink then passes it to the process controller whose job is to control the process variable under some required value. The receiver is responsible for the sensor network management and is controlled and managed by the host application management. The Network and Security Manager is responsible for entire network monitoring and ensuring security against attacks. Therefore, WSN has the potential to improve production processes and quality of products without compromising the IIoT QoS. Actuation and control, and sensing, are also imperative in majority industrial applications. In these applications, the sensors detect the data and the actuators act on the data based on certain control decisions made by the process controller.

4.2 Federated learning implementation challenges in IIoT and IWSN

To implement FL's full potential in IIoT and IWSN, there are still several fundamental challenges that need to be addressed. In this section, we describe the challenges followed by very promising opportunities to meet those challenges.

4.2.1 Limited computational resources

Indeed, the FL deployment on IIoT and IWSN networks relies deeply on the computational resources and memory of edge devices. Consequently, people often focus only on the IoT devices capabilities to gather data while ignoring their limited memory and compute resources, which makes it hard for most IoT and sensor devices to finish local computation with massive data or sophisticated models. In order to address this challenge, lightweight AI techniques have been explored, which can be implemented in resource-constrained FL-IoT and WSN environments, such as improved resource management approaches to accelerate FL training on devices.

4.2.2 Device heterogeneity

In the multi-device settings, participants under the FL framework have various system resources, such as compute and memory resources. As the trend in machine learning is for larger and deeper models, the hardware heterogeneity within the IIoT and IWSN systems pose several challenges for the FL structure. They could easily train large models as devices with powerful memory and computing resources while other devices with limited resources could only train smaller models. Reader speed will also vary across devices, even for the same model size, which can trigger the problem of asynchronous communication discussed above. Due to the availability of the resources, an FL framework for IIoT and IWSN should provide a graceful adaptation of data and compute load on diverse devices.

4.2.3 Limited networking bandwidth

Communication overload is considered to be one of the main challenges in FL-based IIoT and IWSN environments. Currently, most IoT and WSN devices communicate using wireless networks that have a much lower bandwidth than the wired network bandwidth. As more and more devices join the system, the communication problem arises when the clients have different resource allocations. The limited network bandwidth not only makes the communication between clients and the server inefficient, but also triggers the presence of late clients, which fail to share their local update with the server during the communication cycle. To meet this challenge, some key ideas can be used, such as decentralized training, data compression and participant selection.

4.2.4 Adversarial attack and defense

The IoT devices prevalence also poses an attractive target in the real-world deployment for adversaries seeking to launch attacks, such as identity theft, phishing, and distributed denial of service (DDoS). Many IoT and ISN devices do not have the compute resources to do so, although these attacks can be easily defended by installing security patches. It is critical for the IIoT andd IWSN systems to detect the malicious or broken IoT devices that will ruin the model training with limited resource. To address this challenge, one of the promising directions is to implement a lightweight security protocol in the IIoT andd IWSN systems for the detection of broken and malicious devices.

4.2.5 Expected future solutions

Undeniably, the IIoT and IWSN ecosystems continue to evolve at a breakneck pace, exceeding all growth expectations and ubiquity barriers. From sensor to cloud, this giant network keeps breaking technological bounds in several domains, and wireless sensor nodes are expected to be predominant as the number of IoT devices grows toward the trillions to connect the unconnected world and things. However, their future in the IIoT and IWSN ecosystems still seems foggy, where several challenges, such as device's connectivity, artificial intelligence (AI) at the edge, security and privacy concerns, growing energy needs, the right technologies to be used and keep pulling in opposite directions. To address these issues, which are caused by the complexity and variability of the environment, advanced computing related technologies are widely applied. However, the edge computing is limited by cost, volume, power consumption, and other conditions, so the capacity of edge computing cannot be fully exploited. So that edge computing fully exploits its characteristics of flexible management, federated and collaborative execution and heterogeneous environment, the reconfigurable real-time computer system based on FPGA SoC is strongly recommended. The system, as depicted in **Figure 3** can be built in real time as needed, by the characteristics of the FPGA SoC, including its reconfigurability, partial and total and precise clock control.

A multi-threaded huge number, computing requirements, and parallel heterogeneous data processing are persistently proposed in many environments of manufacturing. However, depending on the multi-environment's requirements with different multiple tasks and several scenes, a single algorithm can no longer face the

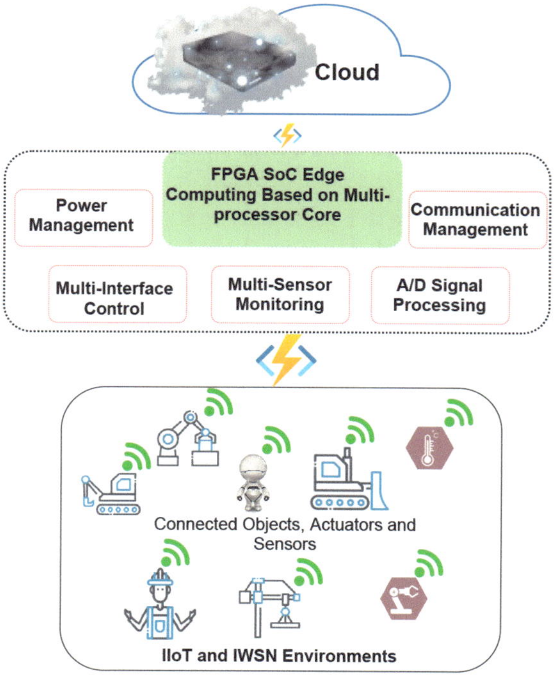

Figure 3.
Reconfigurable edge computing system based on FPGA SoC.

requirements so that numerous complicated tasks require the algorithm to be reconfigured and replaced. Without a doubt, the FPGA employment gratifies this multitude of requirements. However, it can rebuild the logic of the chip by means of configuration and reconfiguration of the resources inside the chip to form hardware with different functions by means of software. Therefore, in addition to the programmability and flexibility of the software, the FPGA also exhibits high throughput, low power, and low latency characteristics. In addition, due to its rich In-Output, FPGA SoCs are also very relevant for use On-chip protocols applications and interface conversion. The main benefits from employing FPGA for the edge computing are as follows:

- A constant throughput can be provided by the FPGA with a constant load size-based application, so that can integrate multiple service requests from several sensors in the IoT.
- Large-scale temporal and spatial parallelism is provided by the FPGA with fine granularity, so that ensures a high concurrency and high dependency algorithm with high acceleration performance.
- Compared with the processor, the FPGA has the lower power consumption and faster computing speed, which can provide the stability and lower task energy consumption.

5. Network slicing architecture and system model

5.1 Network slicing architecture

The 5G network infrastructure design should focus on attentive consideration of software control, hardware infrastructure, and interconnection between them. In this context, we consider a network slicing architecture consisting of a set of IIoT slices $J = \{1, \dots, j\}$, where j represents the slice number. These slices are built on a unified physical infrastructure and share the same network resources. The proposed architecture, which is denoted in **Figure 4**, consists of three virtual slices. The urgent slice is the UCLE which yields more significance to the QoS and the efficiency. Thereafter, the HCLE slice that donates less importance to the latency. The last one is defined as the LCLE, which has the lowest slice priorities with unsecured QoS. The table denoted in the work [6] represents the slice's QoS requirements adopted for our architecture. This architecture consists of a set of $K = \{1, \dots, k\}$ gateways, where k is the number of gateways. Then, gateways take over the task of providing radio resources to the substrate network layer, which contains a set of $I = \{1, \dots, i\}$ IIoT devices, affected to the slice that meets its QoS demands.

5.2 Slicing system model

This slice set $j \in J$ is integrated virtually on a Gateways (GWs) set $k \in K$. However, the physical resources of each GW consist of a set of $C = \{1, \dots, c\}$ channels, in which each one includes a bandwidth $b \in B = \{1, \dots, b\}$. The goal of this work is to provide, for slices member, dynamic channel management based on TP and SF tuning. IN this

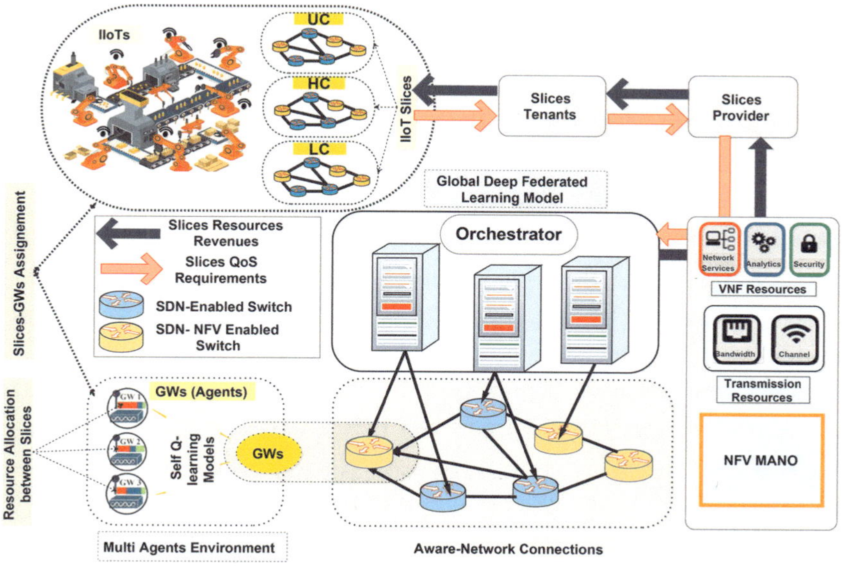

Figure 4.
Deep federated RL-based network slicing architecture.

context, $\alpha_i \in \{0, 1\}$ is denoted as the binary value that indicate the admission success of device $i \in I$ to slice j on GW k. Therefore, we define the Throughput ϕ_i and Delay d_i models, based on SF and TP parameters for each device $i_{j,k}$, as in (1), (2), respectively [7].

$$\phi_i = SF.\frac{R_c}{2^{SF}}.CR = SF.\frac{b_{j,k}}{2^{SF}}.CR, \quad \forall i \in I_{j,k}, \tag{1}$$

$$d_i = \frac{L_i}{\phi_i}, \quad \forall i \in I_{j,k}, \tag{2}$$

where R_c is the chip rate, $b_{i,j}$ denotes the bandwidth assigned for slice j on LoRa GW k, CR represents the coding rate, and L_i is denoted as the packet size. Following the ultimate goal that seek to manage slice's QoS demands, energy efficiency (EE), given in (3), is considered as the second objective that should be maximized for IIoT devices assigned to each slice on each GW.

$$max\, u_{EE}^{j,k} = \sum_{i \in I_{j,k}} \alpha_i \frac{\overline{\phi_i}}{P_j^T + P_c}, \forall i \in I_{j,k}, \tag{3}$$

however, p_i^t denotes the allocation power for each IIoT device. $u_{EE}^{j,k}$ is the EE metric that provides the efficiency of energy efficiency of each slice. While, P_c denotes the power consumption of the circuit and $P_j^T = \sum_{i \in I_{j,k}} p_i^t$ is the TP. Finally, we define the multi-objective problem, as in (4), aiming to maximize the slice utility revenues $U_{rm}^{j,k}$.

$$max\, U_{rm}^{j,k} = \sum_{j,k} \left(u_{QoS}^{j,k} + u_{EE}^{j,k} + u_{REL}^{j,k} \right), \forall k \in K, \forall j \in J, \tag{4}$$

6. The proposed DFRL for network slicing framework

We assume that agents, by sharing its self models based on the Q-network experiences, collaborate to receive global rewards from the federated orchestrator. While the orchestrator collects these models to builds a global network model that provides an optimal actions, on LoRa parameters, that maximize QoS revenue [43, 44].

The considered network consists of two agents, called as agent α and agent β, that play in Markov environment. In this context, we denote the reply memory for agent α by $\mathcal{D}_\alpha = \{s_\alpha, a_\alpha, s'_\alpha, r_\alpha\}$ and the reply memory for agent β by $\mathcal{D}_\beta = \left\{s_\beta, a_\beta, s'_\beta, r_\beta\right\}$. These memories are used to store transitions parameters which will be collected, during interaction, to build an optimal policy (π^*_α and π^*_β). The notations of Q-functions, states, actions, and policy are denoted, respecting to agents α and β, respectively, as $\{Q_\alpha, s_\alpha \in \mathcal{S}, a_\alpha \in \mathcal{A}_\alpha, \pi^*_\alpha\}$ and $\left\{Q_\beta, s_\beta \in \mathcal{S}, a_\beta \in \mathcal{A}_\beta, \pi^*_\beta\right\}$. Thereby, assuming that states spaces (s_α and s_β), transitions parameters ($\mathcal{D}_\alpha$ and $\mathcal{D}_\beta$), and the Q-network functions (Q_α and Q_β) are different for the defined agents α and β. Each agent builds its own Q-network (Q_α or Q_β), and θ (θ_α or θ_β) parameters. These agents interacts with the DFL model with the aim is to build a global federated model that satisfy dynamic slice's QoS demands exploiting local agents experiences Q_α and Q_β.

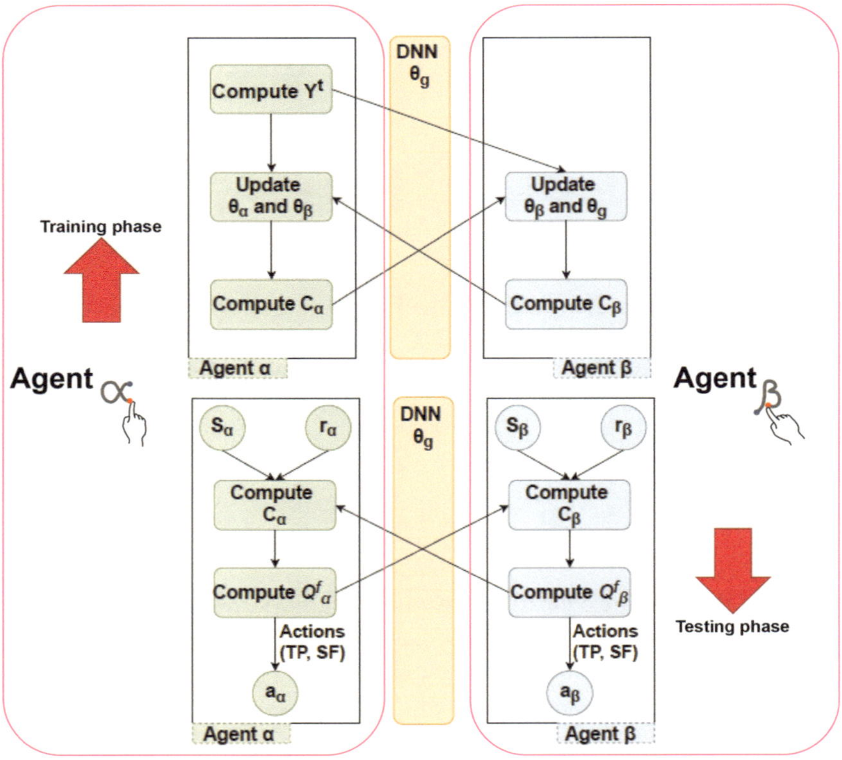

Figure 5.
Training and testing phases.

Therefore, based on the Q-networks models, (5) represents the DFL (based on DNN) Q-network output as $Q_f(\theta_\alpha, \theta_\beta; \theta_g)$.

$$Q_f(\theta_\alpha, \theta_\beta; \theta_g) = DNN\left(\left[Q_\alpha(s_\alpha, a_\alpha; \theta_\alpha) | Q_\beta(s_\beta, a_\beta; \theta_\beta)\right]; \theta_g\right), \tag{5}$$

where $[.|.]$ denotes the concatenation symbol and θ_g denotes the DNN (DFL) parameter shared between agents.

At this stage, the Mean Square Error (MSE), is defined for agents α and β, as a Loss function denoted in formulas (6), (7) [6]. These formulas are used to train the proposed framework, by updating the parameters $(\theta_\alpha, \theta_\beta, \theta_g)$, to build federated model that will be able to find, then, an optimal action decision on TP and SF that maximize slice's QoS rewards.

$$MSE_\alpha^t(\theta_\alpha, \theta_g) = \mathbb{E}\left[\left(Y^t - Q_f^\alpha(s_\alpha^t, a_\alpha^t, C_\beta; \theta_\alpha, \theta_g)\right)^2\right] \tag{6}$$

$$MSE_\beta^t(\theta_\beta, \theta_g) = \mathbb{E}\left[\left(Y^t - Q_f^\beta\left(s_\beta^t, a_\beta^t, C_\alpha; \theta_\beta, \theta_g\right)\right)^2\right], \tag{7}$$

while $Y^t = r^t(s) + \gamma \max_{a \in \mathcal{A}} Q_f^\alpha(s_\alpha^t, a_\alpha^t, C_\beta; \theta_\alpha, \theta_g)$ is attributed for agent α only, as a condition to start training. **Figure 5** depicts the DFRL framework process.

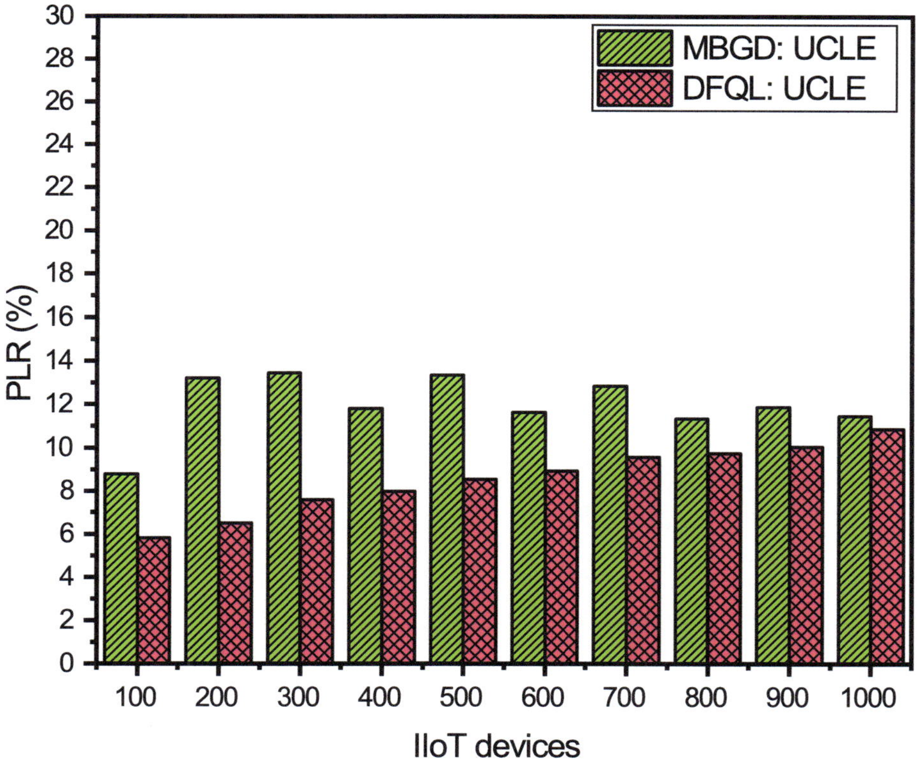

Figure 6.
PLR of UCLE.

7. Experiment results

The proposed framework has been implemented in Python language using TensorFlow-gpu package on Intel Xeon E5-2620 v4 2x 8-Core with 64 GB RAM. Also, the NVIDIA GK110BGL [Tesla K40c] is used to improve speed during the training phase.

We provide, in this section, the mean percentage of Packet Loss Rate (PLR) for IIoT devices, as denoted in **Figures 6–8**, and compare it to the PLR within MBGD scheme [6].

However, by increasing the devices, PLR will increase subsequently. This return to the data rate, that when increase, the number of successful transmitted packet increase accordingly. While it is not the case when throughput is low. We remark also, in **Figures 6–8**, that UCLE and HCLE slices have a reduced PLR compared to LCLE. However, this due to the reliability and efficiency constraints dedicated for this slice which is not the case in LCLE slice that consider only the load. Compared with the slice results using the MBGD technique, we could obviously note the efficiency of the proposed federated scheme in supporting dynamic slicing strategy by reducing PLR over than 9%. This improvement return to the shared experience between agents that can improve the action decision on TP and SF to slices.

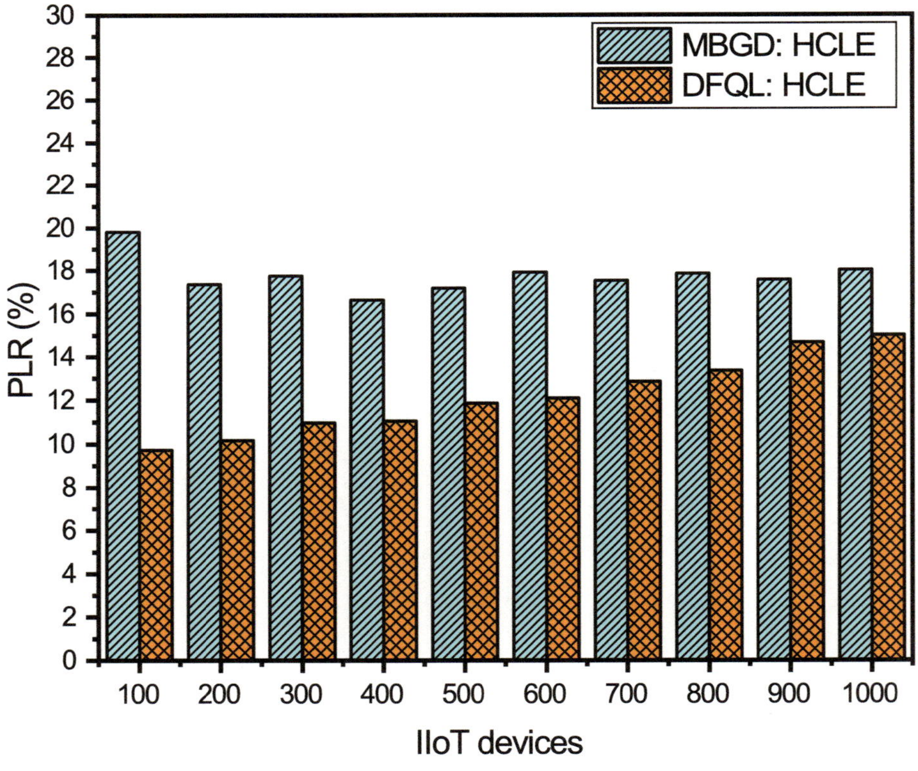

Figure 7.
PLR of HCLE.

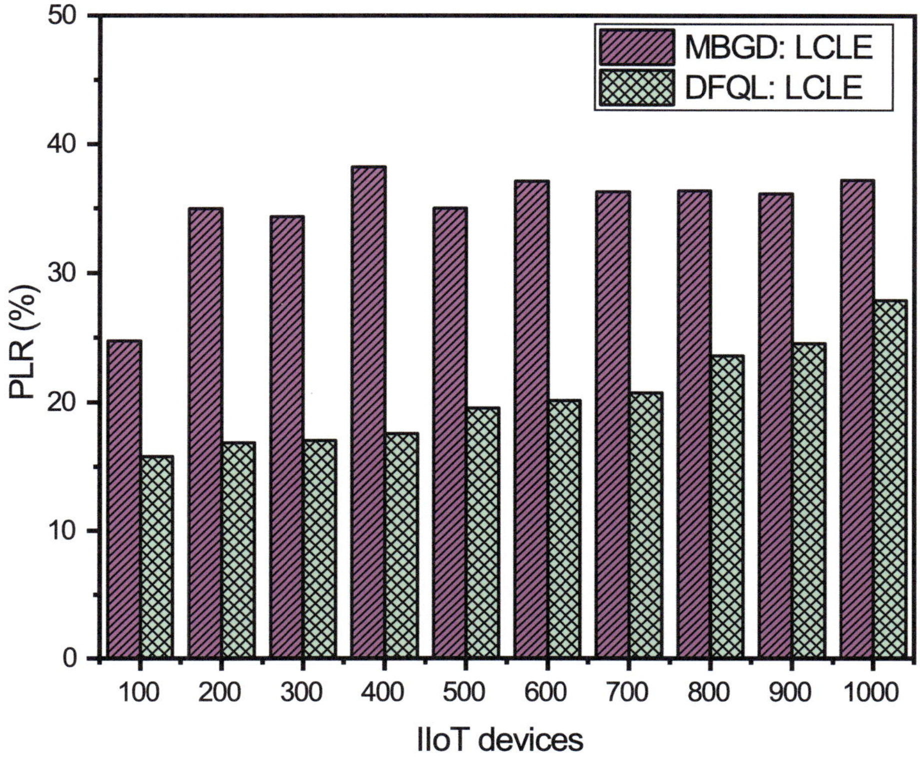

Figure 8.
PLR of LCLE.

8. Conclusion

This prospective chapter presents a future outlook on low-end motes in the IIoT and IWSN eras. Following a detailed discussion of the trends and challenges posed by the IIoT and IWSN paradigm to low-end devices, it discusses how modern reconfigurable platforms are the perfect candidate to meet the ever-evolving industrial environments. Indeed, in this chapter, we proposed a federated network slicing based on deep reinforcement learning techniques for channels and bandwidth management based LoRa promising technology that meet IIoT and IWSN network service requirements based on the SDN, NFV, network slicing, and deep reinforcement learning techniques. Each LoRa GW plays an agent role, in the environment, and profits from the learning experience provided by the other coexist agents via the global federated model.

In the case of future studies, this chapter introduced comprehensive review and several research lines, especially one attractive future line is related to the integration of FPGA SoC at the edge to build a smart factory as well as IIoT and IWSN environments with environmentally friendly capabilities and functionalities. In addition, future research is needed to fully embrace cloud services and new ways of connectivity in order to get the full benefits of the new Edge FPGA SoC technology.

Nomenclature

$J = \{1, \ldots, j\}$	IIoT network slices set
$K = \{1, \ldots, k\}$	LoRa gateways (agents) set
$I = \{1, \ldots, i\}$	IIoT devices set associated to each slice
$B = \{1, \ldots, b\}$	channel bandwidth set
$C = \{1, \ldots, c\}$	LoRa-GW's channels set
$\alpha_i \in \{0, 1\}, \forall i \in I_{j,k}$	device's admission and association index to slice
$\forall i \in I_{j,k}$	device i assigned to slice j on gateway k
TP	transmission power
SF	spreading factor
$\phi_i, \forall i \in I_{j,k}$	throughput of device i
$d_i, \forall i \in I_{j,k}$	delay of device i
$u_{QoS}^{j,k}$	quality of service metric for slice j on GW k
$p_i^t, \forall i \in I_{j,k}$	the power allocated for each device i
$u_{EE}^{j,k}$	energy efficiency metric for slice j on GW k
p_i^r	the received power
$u_{REL}^{j,k}$	reliability metric for slice j on GW k
$U_{rm}^{j,k}, \forall k \in K, \forall j \in J$	the global slice utility revenues metric
$\{\mathcal{S}, \mathcal{A}, \mathcal{T}, \mathscr{R}\}$	state, action, transition function, reward
α and β	agent α and agent β
γ	discount factor
$\theta_\alpha, \theta_\beta$	DQL network parameters (weights)
θ_g	DNN network parameters (weights)
$\mathcal{D}_\alpha, \mathcal{D}_\beta$	reply memories to store transitions

Abbreviations

IoT	internet of things
IIoT	industrial IoT
IWSN	industrial wireless sensor network
AI	artificial intelligence
ML	machine learning
DL	deep learning
QoS	quality of service
RL	reinforcement learning
DRL	deep reinforcement learning
FL	federated learning
DFL	deep federated learning
DFRL	deep federated reinforcement learning
CPIToS	cyber-physical internet of thing systems
5G	fifth generation network
SDN	software defined network
NFV	network function virtualization
NS	network slicing
DQL	deep Q-learning
GD	gradient descent

GMM	gaussian mixture model
SP	service provide
MEC	mobile edge computing
GPU	graphic processor unit
FPGA	field programmable gate array
UCLE	ultra critical of latency and efficiency
HCLE	high critical of latency and efficiency
LCLE	low critical of latency and efficiency
GW	gateway

Author details

Seifeddine Messaoud[1]*, Soulef Bouaafia[1], Abbas Bradai[2], Mohamed Ali Hajjaji[3], Abdellatif Mtibaa[4] and Mohamed Atri[5]

1 Faculty of Sciences, Electronics and Microelectronics Lab., University of Monastir, Monastir, Tunisia

2 XLIM Lab., University of Poitiers, Poitiers, France

3 Electronics and Microelectronics Lab., ISAAT Sousse, University of Sousse, Sousse, Tunisia

4 Electronics and Microelectronics Lab., ENIM Monastir, University of Monastir, Monastir, Tunisia

5 College of Computer Science, King Khalid University, Abha, Saudi Arabia

*Address all correspondence to: seifeddine.messaoud@fsm.rnu.tn

References

[1] Givehchi O, Landsdorf K, Simoens P, Colombo AW. Interoperability for industrial cyber-physical systems: An approach for legacy systems. IEEE Transactions on Industrial Informatics. 2017;**13**(6):3370-3378

[2] Nordrum, A.. Popular Internet of Things. IEEE Spectrum's Technology Blog [Online]. 2016. Available from: http://spectrum.ieee.org/tech-talk/telecom/internet/popular-internet-of-things-forecast-of-50-billion-devices-by-2020-is-outdated

[3] Messaoud S, Bradai A, Bukhari SHR, Qung PTA, Ahmed OB, Atri M. A survey on machine learning in internet of things: Algorithms, strategies, and applications. Internet of Things. 2020;**12**: 100314

[4] Khan LU, Yaqoob I, Tran NH, Kazmi SM, Dang TN, Hong CS. Edge computing enabled smart cities: A comprehensive survey. arXiv preprint arXiv:1909.08747. 2019

[5] Kazmi SMA, Khan LU, Tran NH, Hong CS. Network Slicing for 5G and Beyond Networks. Berlin/Heidelberg, Germany: Springer. 2019;**1**

[6] Messaoud S, Bradai A, Moulay E. Online GMM clustering and mini-batch gradient descent based optimization for industrial IoT 4.0. IEEE Transactions on Industrial Informatics. 2020;**16**(2): 1427-1435

[7] Dawaliby S, Bradai A, Pousset Y. Adaptive dynamic network slicing in LoRa networks. Future Generation Computer Systems. 2019;**98**:697-707

[8] Messaoud S, Bradai A, Atri M. Distributed Q-learning based-decentralized resource allocation for future wireless networks. In: 2020 17th International Multi-Conference on Systems, Signals & Devices (SSD). IEEE; 2020. pp. 892-896

[9] Liang L, Wu Y, Feng G, Jian X, Jia Y. Online auction-based resource allocation for service-oriented network slicing. IEEE Transactions on Vehicular Technology. 2019;**68**(8):8063-8074

[10] Leconte M, Paschos GS, Mertikopoulos P, Kozat UC. A resource allocation framework for network slicing. In: IEEE INFOCOM 2018-IEEE Conference on Computer Communications. Honolulu, HI, USA: IEEE; 2018. pp. 2177-2185

[11] Caballero P, Banchs A, De Veciana G, Costa-Pérez X, Azcorra A. Network slicing for guaranteed rate services: Admission control and resource allocation games. IEEE Transactions on Wireless Communications. 2018;**17**(10): 6419-6432

[12] Guan W, Wen X, Wang L, Lu Z, Shen Y. A service-oriented deployment policy of end-to-end network slicing based on complex network theory. IEEE Access. 2018;**6**:19691-19701

[13] Li D, Hong P, Wang W, Pei J. Virtual network function placement with function decomposition for virtual network slice. In: 2018 IEEE Conference on Standards for Communications and Networking (CSCN). Paris, France: IEEE; 2018. pp. 1-4

[14] Bagaa M, Taleb T, Laghrissi A, Ksentini A, Flinck H. Coalitional game for the creation of efficient virtual core network slices in 5G mobile systems. IEEE Journal on Selected Areas in Communications. 2018;**36**(3): 469-484

[15] Wang G, Feng G, Qin S, Wen R, Sun S. Optimizing network slice dimensioning via resource pricing. IEEE Access. 2019;7:30331-30343

[16] Schneider S, Dräxler S, Karl H. Trade-offs in dynamic resource allocation in network function virtualization. In: 2018 IEEE Globecom Workshops (GC Wkshps). Abu Dhabi, United Arab Emirates: IEEE; 2018. pp. 1-3

[17] Liu J, Shi Y, Zhao L, Cao Y, Sun W, Kato N. Joint placement of controllers and gateways in SDN-enabled 5G-satellite integrated network. IEEE Journal on Selected Areas in Communications. 2018;**36**(2):221-232

[18] Gu B, Feng J, Zhou Z, Guizani M. Time-dependent pricing for on-demand bandwidth slicing in software defined networks. In: 2018 14th International Wireless Communications & Mobile Computing Conference (IWCMC). Limassol, Cyprus: IEEE; 2018. pp. 1024-1029

[19] Wang G, Feng G, Tan W, Qin S, Wen R, Sun S. Resource allocation for network slices in 5G with network resource pricing. In: GLOBECOM 2017-2017 IEEE Global Communications Conference. Singapore: IEEE; 2017. pp. 1-6

[20] Yang K, Jiang T, Shi Y, Ding Z. Federated learning via over-the-air computation. IEEE Transactions on Wireless Communications. 2020;**19**(3): 2022-2035

[21] Yu Z, Hu J, Min G, Lu H, Zhao Z, Wang H, et al. Federated learning based proactive content caching in edge computing. In: 2018 IEEE Global Communications Conference (GLOBECOM). Abu Dhabi, United Arab Emirates: IEEE; 2018. pp. 1-6

[22] Wang S, Tuor T, Salonidis T, Leung KK, Makaya C, He T, et al. When edge meets learning: Adaptive control for resource-constrained distributed machine learning. In: IEEE INFOCOM 2018-IEEE Conference on Computer Communications. Honolulu, HI, USA: IEEE; 2018. pp. 63-71

[23] McMahan HB, Moore E, Ramage D, Arcas BA. Federated learning of deep networks using model averaging. arXiv preprint arXiv:1602.05629. 2016

[24] Smith V, Chiang CK, Sanjabi M, Talwalkar A. Federated multi-task learning. arXiv preprint arXiv: 1705.10467. 2017

[25] Kim H, Park J, Bennis M, Kim SL. Blockchained on-device federated learning. IEEE Communications Letters. 2019;**24**(6):1279-1283

[26] Ren J, Wang H, Hou T, Zheng S, Tang C. Federated learning-based computation offloading optimization in edge computing-supported internet of things. IEEE Access. 2019;7:69194-69201

[27] Zhou Y, Fadlullah ZM, Mao B, Kato N. A deep-learning-based radio resource assignment technique for 5G ultra dense networks. IEEE Network. 2018;**32**(6):28-34

[28] Joung J. Machine learning-based antenna selection in wireless communications. IEEE Communications Letters. 2016;**20**(11):2241-2244

[29] Li H, Ota K, Dong M. Learning IoT in edge: Deep learning for the internet of things with edge computing. IEEE Network. 2018;**32**(1):96-101

[30] Luong NC, Hoang DT, Gong S, Niyato D, Wang P, Liang YC, et al. Applications of deep reinforcement learning in communications and

networking: A survey. IEEE Communication Surveys and Tutorials. 2019;**21**(4):3133-3174

[31] Pham QV, Fang F, Ha VN, Piran MJ, Le M, Le LB, et al. A survey of multi-access edge computing in 5G and beyond: Fundamentals, technology integration, and state-of-the-art. IEEE Access. 2020;**8**:116974-117017

[32] Konečný J, McMahan HB, Yu FX, Richtárik P, Suresh AT, Bacon D. Federated learning: Strategies for improving communication efficiency. arXiv preprint arXiv:1610.05492. 2016

[33] Yang T, Andrew G, Eichner H, Sun H, Li W, Kong N, et al. Applied federated learning: Improving google keyboard query suggestions. arXiv preprint arXiv:1812.02903. 2018

[34] Xu J, Glicksberg BS, Su C, Walker P, Bian J, Wang F. Federated learning for healthcare informatics. Journal of Healthcare Informatics Research. 2021; **5**(1):1-19

[35] Jiang JC, Kantarci B, Oktug S, Soyata T. Federated learning in smart city sensing: Challenges and opportunities. Sensors. 2020;**20**(21): 6230

[36] Sheller MJ, Edwards B, Reina GA, Martin J, Pati S, Kotrotsou A, et al. Federated learning in medicine: Facilitating multi-institutional collaborations without sharing patient data. Scientific Reports. 2020; **10**(1):1-12

[37] Shokri R, Shmatikov V. Privacy-preserving deep learning. In: Proceedings of the 22nd ACM SIGSAC Conference on Computer and Communications Security. Denver, CO, USA; 2015. pp. 1310-1321

[38] Deepa N, Pham QV, Nguyen DC, Bhattacharya S, Prabadevi B, Gadekallu TR, et al. A survey on blockchain for big data: Approaches, opportunities, and future directions. arXiv preprint arXiv:2009.00858. 2020

[39] Hakak S, Khan WZ, Gilkar GA, Assiri B, Alazab M, Bhattacharya S, Reddy GT. Recent advances in blockchain technology: A survey on applications and challenges. arXiv preprint arXiv:2009.05718. 2020

[40] Wang YE, Wei GY, Brooks D. Benchmarking tpu, gpu, and cpu platforms for deep learning. arXiv preprint arXiv:1907.10701. 2019

[41] Cho HD, Engineer PDP, Chung K, Kim T. Benefits of the Big. LITTLE Architecture. EETimes; 2012

[42] Liu L, Han G, Xu Z, Shu L, Martinez-Garcia M, Peng B. Predictive boundary tracking based on motion behavior learning for continuous objects in industrial wireless sensor networks. IEEE Transactions on Mobile Computing. 2021. Early Access

[43] Haque ME, Baroudi U. Ambient self-powered cluster-based wireless sensor networks for industry 4.0 applications. Soft Computing. 2021;**25**(3):1859-1884

[44] Messaoud S, Bradai A, Ahmed OB, Quang PTA, Atri M, Hossain MS. Deep federated q-learning-based network slicing for industrial IoT. IEEE Transactions on Industrial Informatics. 2020;**17**(8):5572-5582

Section 3

Wireless Sensor Networks Management

Chapter 4

Resource Allocation in Wireless Body Area Networks: A Smart City Perspective

Beom-Su Kim, Babar Shah and Ki-Il Kim

Abstract

Healthcare is an essential service in smart cities. To deploy healthcare systems in such cities, personal health monitoring systems, infrastructure for collecting and delivering individual data, and a system for diagnosing symptoms are required. For the first requirement, wireless body area networks (WBANs) have recently received considerable attention from research communities. Owing to their main distinguishable features from general wireless sensor networks, research challenges regarding WBANs have been focused on network topology around the body and implanted nodes, efficient resource allocation, and power control. In this chapter, we provide a comprehensive discussion on the emerging research trends in the area of wireless sensor networks and a discussion of WBANs in terms of their resource allocation.

Keywords: wireless body area networks, resource allocation, radio resource control, transmission power control, smart city

1. Introduction

A smart city can be defined as a converged IT-based infrastructure that can provide information to civilians whenever required, as well as efficiently manage its elements as illustrated in **Figure 1**. As many studies have previously mentioned [1–3], the key technological needs of smart cities include the collection of diverse sensor' data and the monitoring and management of community services. In addition, there are five elements for a smart city based on the layer concept:

1. hardware infrastructure, that is, physical components, such as transportation and buildings;

2. sensors, that is, sensors and terminal nodes;

3. networks, that is, wired and wireless networks including WiFi and metropolitan area networks;

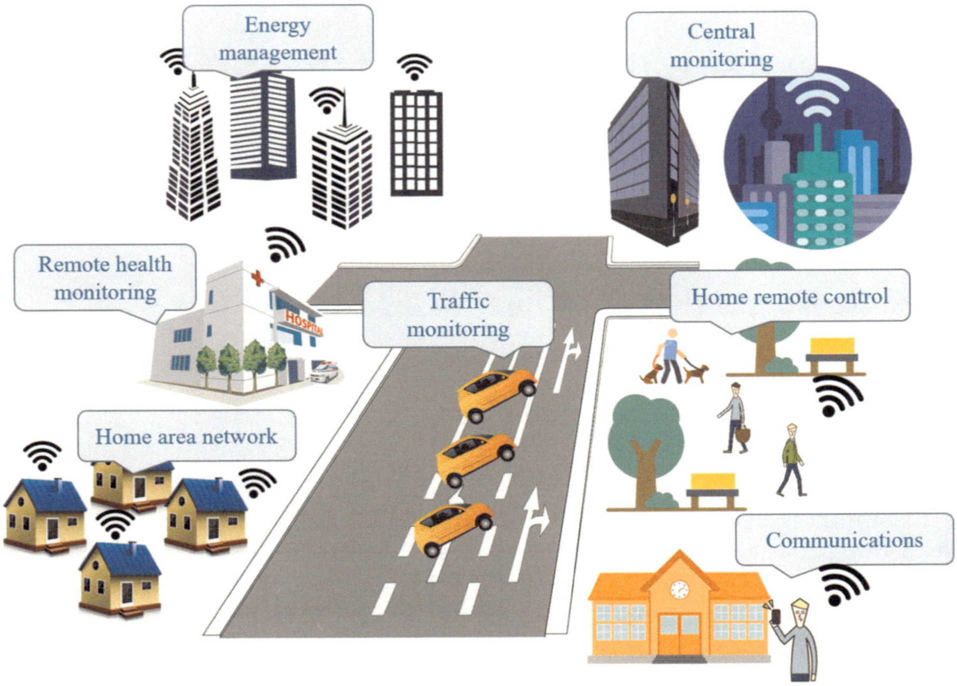

Figure 1.
Illustration of smart city architecture.

4. data and support, that is, application support through data collection, storage, and analysis;

5. applications, service elements, such as a smart economy, healthcare, education, and government.

Especially, applications combined with the Internet of Things (IoT) have been consistently mentioned in the smart city. They include city lighting, city transit, environmental and wastewater management, and privacy-preserving. Among these important services, healthcare is essential for efficiently collecting patient information and diagnosing any symptoms in a proactive way, as addressed in refs. [4, 5]. In addition, prompt treatment is also available from doctors without a direct examination through healthcare applications.

To deploy these services in a smart city, a personal health monitoring system and infrastructure networks are required. For the former, a new type of wireless sensor network called a wireless body area network (WBAN) has been proposed. In addition, IEEE 802.15.6 [6] has been established as an international standard for a WBAN.

In the aspects of telecommunications technology, WBAN is one of the special types of general wireless sensor networks in that wearable and implantable sensor nodes are supposed to detect and reports pre-determined events toward the sink node, called coordinator. In typical sensor networks, nodes are usually spatially distributed in the target area with a communications interface. This implies that they build the networks in an autonomous way. These tiny sensors naturally have resource problems, such as power consumption, limited computing capability, communication range, and available bandwidth. To protect the limitation of a sensor node, efficient and smart usage of available resources is more important than typical wireless networks.

Although a WBAN can be considered a type of wireless sensor network, its main features are significantly different from those of a general wireless sensor network (WSN) in that the sensor nodes are implanted or deployed on the body. This implies that a more careful network design should be taken such that the sensor nodes cannot harm the tissue of an individual through an increase in the temperature. In addition, more severe conditions for operations in WBANs require intelligent resource allocation to efficiently monitor and collect patients' health information. The main conditions to be considered for resource allocation in a WBAN are as follows:

1. Body movement: In a WBAN, the sensors are attached to the surface of, or inserted inside, the body; thus, body movement causes significant channel fading between the coordinator and sensor. Channel fading from body movement is referred to as body shadowing, and a device that aggregates physical data and transmits them to a medical server is called a coordinator. Because body shadowing causes packet loss, the link quality between the coordinator and the sensor is a key factor that should be considered for resource allocation.

2. Heterogeneous sensor types: Depending on the user's health condition, a WBAN can be composed of various types of medical sensors. In addition, the IEEE 802.15.6 standard specifies device priorities from 0 to 7 according to the application type. This means that a differentiated quality of service (QoS) must be guaranteed for a node collecting important body data; hence, resource allocation techniques must consider device priorities.

3. Non-rechargeable battery: Because most sensors are inserted into the body, it is difficult to replace or recharge their batteries. To extend the lifetime of a node, it is necessary to minimize the duty cycle or reduce resource waste owing to retransmission.

These conditions are the key factors distinguishing a WBAN from general sensor networks and have led to the development of network architecture and resource allocation techniques specialized for WBANs. In this chapter, we provide an overview of network architectures for WBANs and their design strategies. Next, we survey resource allocation techniques for WBANs and classify them into two categories—radio resource control and transmission power control. By clarifying their operating mechanisms and research objectives, we provide a comprehensive research trend and discussion of WBANs in terms of their resource allocation.

2. Overview of IEEE 802.15.6 based WBANs

In general, WBANs are specific networks for personal health monitoring, as illustrated in **Figure 2**. It is called intra-WBAN by being distinguished from inter-WBANs. The coordinator is located at the center of the body and is interconnected with the sensor nodes in a one-hop star topology. The coordinator aggregates the data received from the nodes and transmits them to the medical server through an external wireless network.

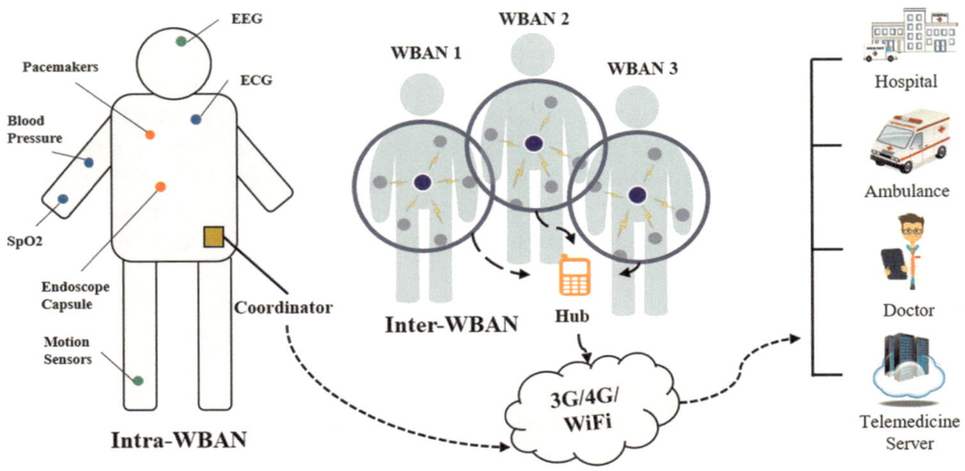

Figure 2.
Illustration of WBAN architecture.

The coordinator allocates available resources to nodes through centralized control. To manage the available resources, the IEEE 802.15.6 standard provides unique resource allocation mechanisms, such as association, access mode, and access phase. In this section, we provide an overview of the major components of the IEEE 802.15.6 standard used for resource allocation.

2.1 Access mode

According to the IEEE 802.15.6 standard, the coordinator operates in one of three access modes—beacon mode with superframes, non-beacon mode with superframes, and non-beacon mode without superframes. The coordinator can select an appropriate access mode considering the application requirements, channel conditions, and policy regulations to save available resources. It should be noted that the resource allocation techniques described in this chapter adopt beacon mode with superframes as the channel access mechanism.

2.2 Access phase

As shown in **Figure 3**, three types of access phases [i.e., random access phase (RAP), exclusive access phase (EAP) and managed access phase (MAP)] can be arranged in the superframe. Each node contends for channel acquisition in the RAP and EAP using carrier-sense multiple-access with collision avoidance (CSMA/CA). The contention window (CW) boundary in CSMA/CA is determined by a predefined value between and based on the device priority. As given in **Table 1**, the device

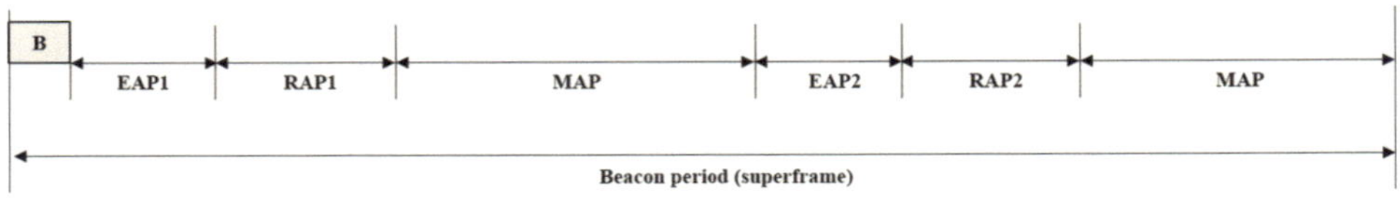

Figure 3.
Beacon mode with superframes.

Priority (0–7)	CW_{min}	CW_{max}
0	1	4
1	2	8
2	4	8
3	4	16
4	8	16
5	8	32
6	16	32
7	16	64

Table 1.
Contention window bounds for CSMA/CA specified in the IEEE 802.15.6 standard.

priorities for CSMA/CA are divided into eight levels (i.e., 0–7), giving a higher value to a node that collects important body data.

In RAP, all nodes are accessible, whereas only high-priority nodes can access EAP. In MAP, the coordinator can contain scheduled uplink allocation intervals, scheduled downlink allocation intervals, and scheduled bi-link allocation intervals. To obtain a scheduled allocation interval, each node must establish an association with the coordinator. The association process is described in the following subsections.

2.3 Connection-oriented resource allocation

The IEEE 802.15.6 standard specifies that each sensor must notify its QoS requirements through an association with the coordinator. To establish an association, an unconnected node sends a connection request frame to the coordinator. The connection request frame includes the device priority and the requested number of timeslots. Upon receiving the connection request frame, the coordinator conducts timeslot scheduling and then notifies the scheduling information through a connection assignment frame.

After establishing an association, each node sends a data frame to the coordinator with the default output power. To change the scheduling information or adjust the transmission power level after the initial association, the coordinator unicasts the changed information to the corresponding node through a new connection assignment frame. Because the IEEE 802.15.6 standard does not define a specific header for transmission power control, the protocol designer must define a reserved space in the connection assignment frame to notify the transmission power level.

2.4 Acknowledgment policy

The IEEE 802.15.6 MAC supports two types of ACK policies to achieve energy-saving and scheduling efficiency. A source node can set the ACK policy field of the control frame. A receiver sends an immediate acknowledgment (I-ACK) or block acknowledgment (B-ACK) frame when it receives a data frame. For example, if the coordinator adopts the B-ACK mode, the control overhead and transmission delay for a continuous data stream can be reduced, whereas, in a situation in which packet loss owing to body shadowing frequently occurs, the I-ACK mode can improve the transmission reliability.

3. Resource allocation in WBANs

As described in the previous section, the IEEE 802.15.6 standard specifies the network architecture, access mode, and frame structure for WBANs; however, it does not define specific algorithms for radio resource control and transmit power control. To complement the IEEE 802.15.6 standard, various resource allocation techniques based on the IEEE 802.15.6 standard have been proposed. As previously described, they aim to satisfy the inherent constraints of a WBAN, such as body movement, heterogeneous sensor types, and non-rechargeable batteries. In this section, we classify the resource allocation techniques into radio resource control and transmission power control and then describe their operating mechanisms.

3.1 Radio resource control

In beacon mode with superframes, the coordinator can use time-division multiple-access (TDMA) or CSMA/CA as a channel access mechanism. In this subsection, we introduce TDMA-, CSMA/CA-, and hybrid-based approaches for allocating radio resources to sensor nodes.

3.1.1 TDMA-based approaches

Alam et al. [7] proposed an adaptive scheduling scheme to reduce idle energy consumption. To minimize energy waste through idle listening and overhearing, the coordinator creates a traffic register and records the sampling interval of the nodes. The coordinator then allocates a timeslot based on the sampling interval of the nodes using the traffic register. Each node can minimize the consumption of idle energy by waking up according to its own wake-up schedule.

In addition, Zhang et al. [8] proposed a data-rate-aware scheduling algorithm to improve energy efficiency. The authors pointed out that when allocating more timeslots to nodes with high data rates, a differentiated QoS can be supported; however, the energy is quickly consumed. To deal with this problem, the coordinator allocates the same number of timeslots to all nodes but adds additional timeslots to high-data-rate nodes when abnormal conditions are detected.

Ambigavathi et al. [9] proposed a priority-based scheduling technique to minimize delays in emergency data. Basically, the coordinator allocates timeslots in the order of device priority. The coordinator divides the data received into low- and high-threshold data to guarantee a differentiated QoS for emergency data occurring at runtime. The coordinator then preferentially allocates timeslots to a node with a higher criticality among nodes with the same device priority.

Liu et al. [10] also proposed a dynamic scheduling technique to reduce the packet loss owing to body movement. The authors used a Markov decision-making process to recognize body movement. The proposed Markov model is defined in two states, that is, a "good state" and a "bad state." Each state is determined based on the link quality between the coordinator and the sensor. For example, a good state indicates that the transmission has been successful, and a bad state indicates that the transmission has failed. For each TDMA round, the coordinator preferentially allocates timeslots to nodes in a good state.

Zhang et al. [11] proposed a dynamic scheduling technique using temporal autocorrelation to reduce the packet loss owing to body shadowing. The authors pointed out that when defining the state of an on-body link as good or bad, the channel condition cannot be accurately recognized. To overcome this limitation, the proposed scheme predicts the channel condition in the next TDMA round after analyzing the temporal autocorrelation using historical data on the link quality. The coordinator then preferentially allocates timeslots to nodes with high autocorrelation coefficients because the higher the autocorrelation coefficient, the more uniform the movement pattern.

Kim et al. [12] proposed a scheduling order optimization technique using multiple cognitive metrics to achieve multi-objective optimization, such as a differentiated QoS and energy saving. The authors used the multi-criteria decision-making (MCDM) method [13] to combine the three cognitive metrics (i.e., packet error rate, power consumption ratio, and user priority) into a single metric called the critical index. Specifically, the proposed MCDM model derives a weighted normalized value (i.e., critical index) after determining the relative importance of multiple metrics through a pairwise comparison matrix. As shown in **Figure 4**, the pairwise comparison matrix should be determined in advance by the decision-maker based on the network conditions. The coordinator then ranks the nodes based on the critical index and determines the scheduling order. A similar study was conducted by Roy et al. 1 [14].

Pushpan et al. [15] proposed an adaptive scheduling scheme to support energy efficiency and a differentiated QoS. As illustrated in **Figure 5**, the authors used fuzzy logic [16] to unify multiple cognitive metrics. The proposed fuzzy model uses the packet delivery ratio, energy consumption ratio, and buffer ratio as input values and normalizes them using linguistic terms (e.g., low, medium, and high). After defining a fuzzy inference rule through an if-else conditional statement, a fuzzy index can be derived. For example, a node with a "high" packet delivery ratio and "low" energy consumption ratio may acquire a higher fuzzy index using the fuzzy inference rule. The coordinator then determines the scheduling order based on the fuzzy index. Similar concepts can be found in refs. [17–19].

Chowdhury et al. [20] proposed a dynamic scheduling scheme to jointly improve the energy efficiency and network QoS. The authors derived the optimal scheduling policy through trial and error using Q-learning. The coordinator (i.e., agent) defines the state space as a combination of the sum rate and response time and the action

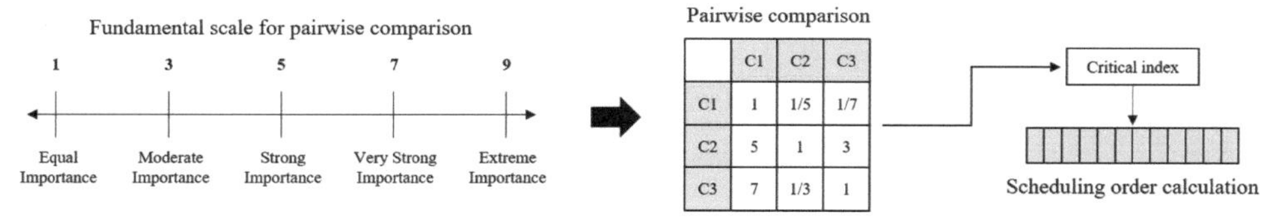

Figure 4.
Scheduling order calculation using the MCDM method.

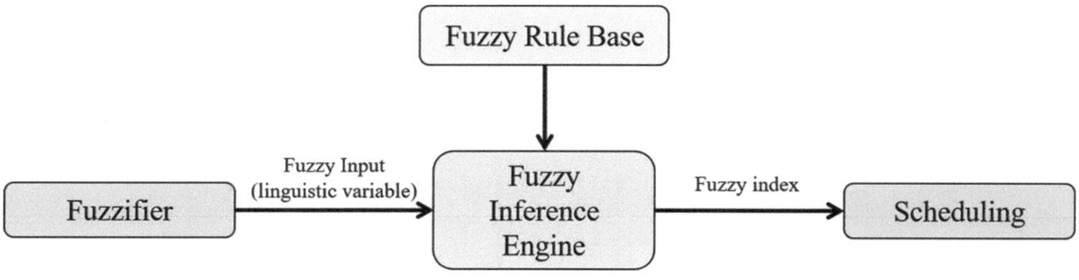

Figure 5.
Scheduling-order calculation using fuzzy logic.

space as the number of timeslots. The reward function is defined as a combination of the sum rate, response time, and average delay. The coordinator then finds a scheduling policy that maximizes the reward function while randomly allocating timeslots to nodes based on the epsilon-greedy algorithm (**Table 2**).

Chen et al. [21] proposed a joint optimization scheme using a deep Q-network (DQN). When a state space is defined through a combination of cognitive metrics, the coordinator has numerous state-action combinations. Thus, the scheduling technique using Q-learning has a problem in that the volume of the Q-table increases. To solve this problem, as shown in **Figure 6**, the authors used the DQN to learn the optimal scheduling policy. The proposed DQN model defines a state space as a combination of device priority, battery level, average delay, link quality, and access time as an action space. The reward function is defined as a combination of the energy consumption ratio, average delay, and received signal strength. The coordinator creates transitions using Q-learning and stores them in the replay memory. The coordinator then trains the neural network by randomly sampling transitions from this memory. Similar concepts can be found in refs. [22, 23].

3.1.2 CSMA/CA-based approaches

Saboor et al. [24] proposed a dynamic backoff algorithm for increasing the superframe utilization and energy efficiency. A typical binary exponential backoff

Reference	Research objective	Major consideration	Cognitive metric
[7]	Energy-saving	Heterogeneous traffic flow	Sampling rate
[8]	Energy-saving	Heterogeneous traffic flow	Abnormal condition
[9]	Differentiated QoS	Heterogeneous traffic flow	Traffic criticality
[10]	Energy-saving	Body shadowing	On-body link quality
[11]	Transmission reliability	Body shadowing	On-body link quality
[12]	Energy efficiency, differentiated QoS	Heterogeneous traffic flow, body shadowing	Packet error rate, priority, power consumption ratio
[15]	Energy efficiency, differentiated QoS	Heterogeneous traffic flow, body shadowing	Packet delivery ratio, energy ratio, buffer ratio
[20]	Energy efficiency, network QoS	Heterogeneous traffic flow, body shadowing	Sum rate, response time
[21]	Energy efficiency, differentiated QoS	Heterogeneous traffic flow, body shadowing	priority, battery level, delay

Table 2.
Summary of TDMA-based radio resource control.

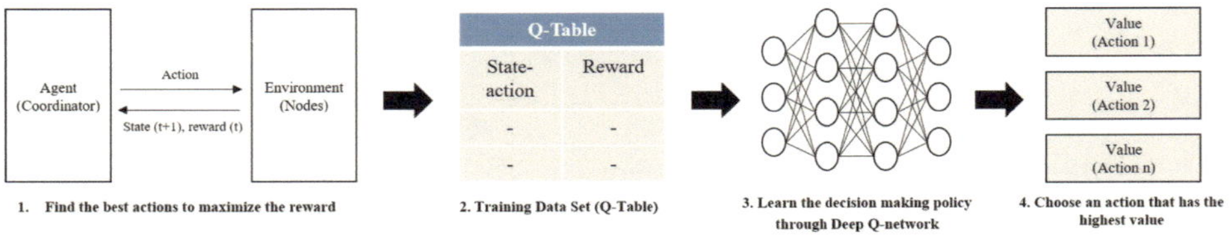

Figure 6.
DQN training for time-slot scheduling.

mechanism doubles the CW size when an even number of collisions occur; however, this mechanism reduces the superframe utilization. To solve this problem, the proposed backoff algorithm employs an additional sliding window. Specifically, the backoff number is randomly initialized within the range given in **Table 1**. If the channel is idle, the backoff number is decreased by 1, and when it becomes zero, the node transmits a data frame. If an even number of collisions occur, instead of doubling, it is replaced with the value of the sliding window. Here, the value of the sliding window was set to a non-overlapping value between the nodes. By using non-overlapping contention windows between nodes, the collision probability is reduced and superframe utilization is increased.

In addition, Fourati et al. [25] proposed a dynamic backoff algorithm to support a differentiated QoS. In a binary exponential backoff mechanism using the given CW size shown in **Table 1**, low-priority nodes have a longer waiting time when an even number of collisions occur. To solve this problem, the proposed algorithm uses a starvation index and continuously changes the CW boundary at runtime. Specifically, the starvation index is initialized to zero and increased by 1 when the channel is busy or retransmission occurs. If an even number of collisions occur, the CW boundary is recalculated based on the starvation index. By estimating the latency of a given packet, the proposed scheme can maintain a balance between low priority and critical traffic.

Mouzehkesh et al. [18] also proposed a dynamic backoff algorithm using fuzzy logic to improve the network reliability and overall latency. To ensure a balance between the channel condition and waiting time, the proposed fuzzy model uses the channel clear rate (CCR) and sample rate (SR) as input variables. The fuzzy inference rule gives a higher fuzzy index to a node having a "high" CCR and a "medium" SR. The backoff delay is then determined based on the fuzzy index.

Nekooei et al. [19] proposed a dynamic backoff exponent algorithm using fuzzy logic to increase the network reliability. The proposed algorithm represents the busyness of the channel as a backoff rate and calculates the backoff exponent using a combination of the history of the channel condition and the data rate of the node. Specifically, the backoff rate and data rate are used as the input variables for the fuzzy model. The fuzzy index is divided into four levels, and a node with the lowest fuzzy level has a relatively high probability of acquiring a channel (**Table 3**).

3.1.3 Hybrid-based approaches

Contention-free and contention-based approaches aim to improve the efficiency of radio resource control; however, their mechanisms have clear advantages and disadvantages. For example, although TDMA-based approaches can easily support a differentiated QoS, they cannot cope with packet loss owing to body shadowing. By

Reference	Research objective	Major consideration	Cognitive metric
[24]	Energy-saving	Heterogeneous traffic flow	Traffic priority
[25]	Differentiated QoS	Heterogeneous traffic flow	Starvation index
[18]	Transmission reliability	Heterogeneous traffic flow	Channel clear rate, sampling rate
[19]	Transmission reliability	Body shadowing	Number of backoff rate, data rate

Table 3.
Summary of CSMA/CA-based radio resource control.

contrast, CSMA/CA-based approaches do not guarantee a differentiated QoS, although they can increase the utilization of superframes. To overcome the limitations of individual approaches, hybrid-based approaches have been proposed.

Ramachandran et al. [26] proposed a link-quality-aware hybrid MAC protocol. The authors argue that body shadowing itself occurs instantaneously; however, the effectiveness of shadowing increases when the body is active for a long period of time. This effect weakens the communication between the implantable sensor nodes, increasing the likelihood of missing life-critical data. In the proposed scheme, the nodes are classified as high- and low-priority nodes. The superframe is divided into three parts: EAP, RAP, and MAP. In MAP, the coordinator decides to apply TDMA to high-priority nodes, whereas low-priority nodes use EAP and RAP. The proposed scheme uses the received signal strength indicator (RSSI) and packet delivery ratio to predict the dynamics of human activity to improve reliability and energy efficiency. In a TDMA period, the coordinator preferentially allocates a timeslot to a node that has a high packet delivery ratio and moves periodically. In addition, long-term body shadowing increases the chances of missing life-threatening medical data and can increase latency. The authors argue that most medical sensors are unable to significantly increase the output power, and thus temporarily using a relay node may be the optimal solution to overcome this situation. In the proposed scheme, the relay node is used when a low packet delivery ratio is detected; otherwise, it is disabled. This relaying mechanism reduces unnecessary energy waste.

Choi et al. [27] proposed an energy-efficient hybrid MAC protocol. The superframe structure of the IEEE 802.15.6 standard consists of EAP, RAP, and MAP. The authors pointed out that the nodes are always active in MAP, resulting in constant energy consumption. To solve the energy waste problem, the proposed mechanism aims to minimize the MAP length. Specifically, EAP2 and RAP2 are arranged after EAP1 and RAP1, and MAP is arranged at the end of the superframe. If a node finishes its transmission in EAP1, it can go into sleep mode to save energy. Each node provides transmission information to the coordinator such that it can synchronize with all nodes.

Wang et al. [28] proposed a dynamic MAC protocol. With the proposed mechanism, the superframe is divided into RAP and MAP, and the lengths of the two phases are dynamically adjusted according to the data rate and data type. Note that CSMA/CA is used in RAP, whereas TDMA is used in MAP. The coordinator allocates timeslots to nodes in the MAP based on the device priority; that is, high-priority nodes have more timeslots. Nodes with no data to send in the buffer save energy by entering a sleep state. Specifically, three priority levels are defined, from 0 to 2, depending on the criticality. A node that has a high data rate and generates emergency data is given priority 0, whereas normal nodes are given priority 2. Initially, the coordinator gives all nodes the highest priority and then sets the threshold ratio to adjust the priority. If the data rate of a node is greater than the threshold, it is classified as a high-priority node, and the remaining nodes are classified as normal nodes. The threshold value was adjusted dynamically to accommodate heterogeneous traffic. To support a differentiated QoS, the coordinator grants a small range of CWs to the node with the highest priority.

Huq et al. [29] proposed a hybrid MAC protocol to provide a differentiated QoS. The propagation of emergency messages requires high reliability and minimal channel access delays. The authors pointed out that neither RAP nor EAP can be used for emergency traffic because of the potential for data loss owing to channel collisions. To address this issue, the proposed scheme dynamically inserts a listening window within

Reference	Hybrid type	Major consideration	Cognitive metric
[26]	TDMA + CSMA/CA	Heterogeneous traffic flow, body shadowing	Device priority, RSSI, packet delivery ratio
[27]	TDMA + CSMA/CA	Heterogeneous traffic flow	Device priority
[28]	TDMA + CSMA/CA	Heterogeneous traffic flow	Criticality
[29]	TDMA + CSMA/CA	Heterogeneous traffic flow	Device priority

Table 4.
Summary of hybrid-based radio resource control.

a contention-free period (MAP). The frequency of the insertion of the listening window was determined by the minimum delay tolerance. The proposed scheme also uses idle timeslots to insert an additional listening window for emergency traffic without impacting the network throughput. Emergency devices can use guaranteed access periods to send data in the listening window without notifying their QoS requirements to the coordinator. This process can improve the reliability and access times. Note that the duration of the access phase is dynamically adjusted based on the QoS requirements of each node (**Table 4**).

3.2 Transmission power control

Quwaider et al. [30] proposed a body-posture-based transmission power control scheme to strike a balance between energy consumption and reliable transmission. Specifically, the authors proposed a dynamic posture position inference algorithm that recognizes the current posture position using on-body link characteristics. The proposed system infers the current body posture based on RSSI measurements on the receiver side. In addition, the authors set a range of RSSI thresholds to balance a reliable transmission and energy consumption through quantitative experiments based on a real WBAN testbed. The proposed algorithm recognizes the current body posture by defining the relationship between the power-level index and RSSI as a linear equation. Once the linear equation for the new position is obtained, the optimal transmission power level can be derived (**Table 5**).

Zang et al. [31] proposed an accelerometer-assisted transmission power control algorithm to improve energy efficiency. The authors pointed out that energy-efficient communication can be achieved by optimizing the output power required for successful transmission. The existing approaches determine the transmission power level based on the received signal strength; however, there is a possibility that the current link information is already out of date owing to dynamic on-body link conditions. To

Reference	Research objective	Major consideration	Cognitive metric
[30]	Energy-saving, transmission reliability	Body shadowing	Gait-cycle
[31]	Energy-saving, transmission reliability	Body shadowing	Gait-cycle
[32]	Energy-saving, transmission reliability	Body shadowing	On-body link quality
[33]	Energy-saving, transmission reliability	Body shadowing	On-body link quality

Table 5.
Summary of transmission power control techniques.

solve this problem, the authors used periodic link-quality fluctuations. The proposed algorithm defines the relationship between body posture and channel periodicity and then recognizes the current body movement using an accelerometer. This means that the acceleration signal and received signal strength of the packet have the same cycle. Each node then sends a packet with the minimum output power when the link quality is the best or increases the transmission power level using feedback information to prevent a delay violation.

Zhang et al. [32] proposed a joint transmission power control and scheduling scheme to provide a flexible trade-off between transmission reliability and energy consumption. Initially, each node is assigned its own scheduled uplink interval (SUI) with the same number of timeslots to satisfy the fairness constraint. To avoid packet loss from body movement, the authors proposed a temporal autocorrelation model in which the coordinator tracks the link conditions of all sensor nodes based on the RSSI measurements and predicts the channel state in the next TDMA round. The coordinator uses the predicted channel conditions to adjust the SUI order and the transmission power level. For example, the coordinator rearranges the SUI according to the node-link quality to increase the probability of successful transmission. In addition, the output power is determined to be higher than the reception sensitivity, considering the channel fluctuations in the next TDMA round.

Finally, Zhang et al. [33] proposed a relay-aided transmission power control scheme to increase transmission reliability and energy efficiency. The authors pointed out that the transmission power level should be adaptively adjusted to cope with changes in the on-body link conditions. The coordinator recognizes the channel state using the RSSI records of the packets received in the previous TDMA round. The coordinator then calculates the optimal transmission power level required for successful transmission based on the channel state and informs the transmission power level to nodes through a beacon frame. In addition, the authors proposed an adaptive transmission scheme using a relay node to improve the reliability of transmission. If the current channel condition between the coordinator and the source node is expected to be bad, the coordinator notifies the source node to apply a relay-assisted two-hop transmission.

4. Research challenges

In this section, we present further research challenges and open issues in resource allocation in WBANs.

4.1 Advanced/smart resource management for multi-objective optimization

This chapter introduced various resource management techniques to achieve specific objectives (i.e., differentiated QoS, reliability, and energy-saving). Most approaches aim to achieve a single objective; however, WBANs include serious restrictions different from general WSNs and thus must satisfy different service requirements simultaneously. Therefore, it is important to provide flexible trade-offs between optimization criteria rather than achieving a single-objective optimization. That is, advanced/smart resource management techniques that can adaptively make decisions according to changes in network conditions are required to achieve multi-objective optimization.

4.2 Reliable simulation systems

Various resource management techniques have been proposed to satisfy the service requirements of WBANs; however, the development of simulation systems to verify their performance has not been adequately followed. Although some simulation tools (e.g., OMNet++, OPNET, and NS-3) for building a WBAN environment have been proposed, they do not provide the necessary middleware or framework to import the latest technologies (e.g., DRL). The lack of a simulation system is a major factor in reducing the reliability of modern resource management schemes; hence, the development of reliable simulation systems remains one of the challenging tasks in WBANs.

4.3 Deployment for real WBAN system

To satisfy the service requirements of WBANs, various resource management techniques have been proposed in intra-WBANs. However, for the proposed mechanisms to be applied to an actual WBAN system, it is necessary to solve the mutual interference problem between adjacent WBANs in a public place. In addition, a unified network architecture conforming to the IEEE 802.15.6 standard is required to improve the scalability of the resource management techniques; however, most resource management schemes did not consider the IEEE 802.15.6 standard. Hence, bridging the performance gap between a simulation system and a real WBAN system by fully complying with the specifications presented in the IEEE 802.15.6 standard is one of the major challenges in WBANs.

4.4 Security

It is essential to provide high-level security services because WBANs collect physiological data of the human body. The IEEE 802.15.6 standard specifies encryption algorithms (e.g., DES and AES) that can be used at the MAC level, and thus protocol designers should implement a security model to protect private data. However, these encryption algorithms are outdated to defend against various types of attacks. Therefore, it is necessary to develop a simple and effective security model that can respond to various types of attacks considering the limited resources of WBANs.

5. Conclusions

A WBAN has unique characteristics that distinguish it from general wireless sensor networks, which has led to the development of new network architecture and resource allocation technique. In particular, resource allocation techniques have become increasingly intelligent to meet the significant constraints of WBANs. In this chapter, we provided comprehensive research trends and a discussion of WBANs in terms of resource allocation. Specifically, we first introduced the network architecture of a WBAN and outlined the major components of the IEEE 802.15.6 standard used for resource allocation. Next, we classified the resource allocation techniques into radio resource control and transmission power control and then described their operating mechanisms.

Acknowledgements

This research was supported by an Institute for Information & Communications Technology Planning & Evaluation (IITP) grant funded by the Korean government (MSIT) (No. 2019-0-01343, Training Key Talents in Industrial Convergence Security), and Research Cluster Project, R20143, by Zayed University Research Office.

Conflict of interest

The authors declare no conflicts of interest.

Author details

Beom-Su Kim[1], Babar Shah[2] and Ki-Il Kim[1*]

1 Department of Computer Science and Engineering, Chungnam National University, Daejeon, Republic of Korea

2 College of Technological Innovation, Zayed University, Abu Dhabi, UAE

*Address all correspondence to: kikim@cnu.ac.kr

References

[1] Salahuddin MA, Jahed HS, Abdallah K, Issa K, Mohsen G, Ammar G, et al. Smart cities: A survey on data management, security, and enabling technologies. IEEE Communications Surveys & Tutorials. 2017;**19**(4):2456-2501

[2] Ondrej K, Attila K, Ayca K, Tasgetirenm MF. Future trends and current state of smart city concepts: A survey. IEEE Access. 2020;**8**: 86448-86467

[3] Feng Y, Ming Y, Hong L, Zhao T, Basodi S. A survey on algorithms for intelligent computing and smart city applications. Big Data Mining and Analytics. 2021;**4**(3):155-172

[4] Glen D, Gina S, Cook Diane J, Fritz RL. Using smart city technology to make healthcare smarter. Proceedings of the IEEE. 2018;**106**(4):708-722

[5] Pi-Cheng H, Yuan-Yao S, Pang A-C. A data parasitizing scheme for effective health monitoring in wireless body area networks. IEEE Transactions on Mobile Computing. 2018;**18**(1):13-27

[6] IEEE Communication Society, IEEE Standards for Local and Metropolitan Area Networks–Part 15.6: Wireless Body Area Networks. New York, USA: IEEE Standard 802.15.6–2012; 2012. pp. 1-271

[7] Alam MM, Hamida EB, Berder O, Menard D, Sentieys O. A heuristic self-adaptive medium access control for resource-constrained WBAN systems. IEEE Access. 2016;**4**: 1287-1300

[8] Zhang C, Wang Y, Liang Y, Shu M, Zhang J, Ni L. Low duty-cycling mac protocol for low data-rate medical wireless body area networks. Sensors. 2017;**17**(5):1134

[9] Ambigavathi M, Sridharan D. Traffic priority based channel assignment technique for critical data transmission in wireless body area network. Journal of Medical Systems. 2018;**42**(11):206

[10] Liu B, Yan Z, Chen CW. Medium access control for wireless body area networks with QoS provisioning and energy efficient design. IEEE Transactions on Mobile Computing. 2016;**16**(2):422-434

[11] Zhang H, Safaei F, et al. Channel autocorrelation-based dynamic slot scheduling for body area networks. EURASIP Journal on Wireless Communications and Networking. 2018; **2018**(1):246

[12] Kim B-S, Kim K-I. A priority-based dynamic link scheduling algorithm using multi-criteria decision making in wireless body area networks. In: 2020 28th International Symposium on Modeling, Analysis, and Simulation of Computer and Telecommunication Systems (MASCOTS). Nice, France: IEEE; 2020. pp. 1-8

[13] Triantaphyllou E. Multi-criteria decision making methods. In: Multi-Criteria Decision Making Methods: A Comparative Study. Berlin, Germany: Springer; 2000. pp. 5-21

[14] Roy S, Mallik I, Poddar A, Moulik S. PAG-MAC: Prioritized allocation of GTSS in IEEE 802.15. 4 MAC protocol-a dynamic approach based on analytic hierarchy process. In: 2017 14th IEEE India Council International Conference (INDICON). Roorkee, India: IEEE; 2017. pp. 1-6

[15] Pushpan S, Velusamy B. Fuzzy-based dynamic time slot allocation for wireless body area networks. Sensors. 2019;**19**(9):2112

[16] Sugeno M, Asai K, Terano T. Fuzzy Systems Theory and its Applications. Osaka, Japan: Tokyo Institute of Technology; 1992

[17] Nekooei SM, Chen G. Cooperative coevolution design of multilevel fuzzy logic controllers for media access control in wireless body area networks. IEEE Transactions on Emerging Topics in Computational Intelligence. 2018;**4**(3): 336-350

[18] Mouzehkesh N, Zia T, Shafigh S, Zheng L. D2 MAC: Dynamic delayed medium access control (MAC) protocol with fuzzy technique for wireless body area networks. In: 2013 IEEE International Conference on Body Sensor Networks. Cambridge, UK: IEEE; 2013. pp. 1-6

[19] Nekooei SM, Chen G, Rayudu RK. A fuzzy logic based cross-layer mechanism for medium access control in wban. In: 2015 IEEE 26th Annual International Symposium on Personal, Indoor, and Mobile Radio Communications (PIMRC). Hong Kong, China: IEEE; 2015. pp. 1094-1099

[20] Chowdhury A, Raut SA. A QoS alert scheduling based on q-learning for medical wireless body area network. In: 2018 International Conference on Bioinformatics and Systems Biology (BSB). Allahabad, India: IEEE; 2018. pp. 53-57

[21] Chen G, Zhan Y, Sheng G, Xiao L, Wang Y. Reinforcement learning-based sensor access control for wbans. IEEE Access. 2018;7:8483-8494

[22] George EM, Jacob L. Interference mitigation for coexisting wireless body area networks: Distributed learning solutions. IEEE Access. 2020;**8**: 24209-24218

[23] Wang L, Zhang G, Li J, Lin G. Joint optimization of power control and time slot allocation for wireless body area networks via deep reinforcement learning. Wireless Networks. 2020;**26**: 4507-4516

[24] Saboor A, Ahmad R, Ahmed W, Kiani AK, Alam MM, Kuusik A, et al. Dynamic slot allocation using non overlapping backoff algorithm in IEEE 802.15.6 WBAN. IEEE Sensors Journal. 2020;**20**(18):10862-10875

[25] Fourati H, Idoudi H, Saidane LA. Intelligent slots allocation for dynamic differentiation in IEEE 802.15. 6 CSMA/CA. Ad Hoc Networks. 2018;**72**:27-43

[26] Ramachandran VRK, Havinga PJM, Meratnia N. HACMAC: A reliable human activity-based medium access control for implantable body sensor networks. In: 2016 IEEE 13th International Conference on Wearable and Implantable Body Sensor Networks (BSN). San Francisco, USA: IEEE; 2016. pp. 383-389

[27] Choi JS, Kim JG. An improved MAC protocol for wban through modified frame structure. Int. j. smart home. 2014; **8**(2):65

[28] Jun Wang, Yutian Xie, and Qiong Yi. An all Dynamic MAC Protocol for Wireless Body Area Network. In. 2015 International Conference on Wireless Communications, Networking and Mobile Computing (WiCOM). Shanghai, China: IET; 2015. pp. 25-30

[29] Huq MA, Dutkiewicz E, Fang G, Liu RP, Vesilo R. MEB MAC: Improved channel access scheme for medical emergency traffic in WBAN. In: 2012 International Symposium on

Communications and Information Technologies (ISCIT). Gold Coast, Australia: IEEE; 2012. pp. 371-376

[30] Quwaider M, Rao J, Biswas S. Body-posture-based dynamic link power control in wearable sensor networks. IEEE Communications Magazine. 2010; **48**(7):134-142

[31] Zang W, Li Y. Gait-cycle-driven transmission power control scheme for a wireless body area network. IEEE Journal of Biomedical and Health Informatics. 2017;**22**(3):697-706

[32] Zhang H, Safaei F, Tran LC, et al. Joint transmission power control and relay cooperation for WBAN systems. Sensors. 2018;**18**(12):4283

[33] Zhang Y, Zhang B. A relay-aided transmission power control method in wireless body area networks. IEEE Access. 2017;**5**:8408-8418

Chapter 5

Energy Management in Wireless Sensor Network

Tareq Alhmiedat

Abstract

Usually, wireless sensor networks (WSNs) are installed in large areas to monitor various physical conditions of the environment and forward the collected sensed data to a base station (central node), for instance: gas monitoring, intrusion detection, tracking objects, etc. However, sensor nodes are usually deployed unattended and battery-powered with no external power source. Therefore, WSNs face the challenge of limited energy source available onboard, where packet transmission and sensing functions are the most power consumption factors in WSN. Therefore, to overcome the energy depletion in sensor nodes, it is important to study the energy management issue in WSN. In this chapter, the significance of energy management issue is discussed first, and then the possible energy management strategies for WSN are presented and illustrated.

Keywords: energy management, wireless sensor networks (WSNs), power consumption, energy management strategies, data aggregation, clustering

1. Introduction

A wireless sensor network (WSN) is made up of a set of sensor devices (nodes), which are usually powered by batteries to operate and interconnected through radio links to assure data transmission, processing, and reception. In general, WSNs have a significant potential in different applications in the areas of medical sciences, telecommunications, agriculture, environmental sciences, military services, and surveillance. The increasing demand on the deployment of autonomous sensor nodes and extending the sensor network lifetime can therefore be considered among the main objectives through examining interesting methods and research studies, which optimize the WSN energy consumption, and proposing methods to improve it. These methods can include several action levels that can range from the deployment stage to the information processing and manipulation stage [1].

In general, WSN is a combination of distributed self-governing sensor nodes, which monitor environmental and physical certain conditions, for instance: monitoring humidity, temperature, pressure, etc., and transfer such data through multihop network to the base station. WSN is considered as attractive solutions for many applications in fields [2–9]. **Figure 1** depicts an environmental sensor nodes deployed in a forest area, where sensor nodes may transmit the sensed data through multihop communication to the sink node (base station). The energy capability for the sensor

Figure 1.
Energy management.

nodes allows them to work autonomously and can communicate with other nodes through radio waves through the establishment of the routing mechanisms [10].

Recently, an intensive research has studied and addressed the energy consumption issue in WSNs, as the sensor networks have been employed in various types of applications, where it is difficult in certain cases to replace or recharge the attached battery source. In addition, sensor nodes are expected to work from months to a few years. Therefore, it is significant to develop an energy efficient hardware and software components to allow the WSNs to operate for the maximum period of time. This chapter focuses on the power management issue in WSNs and discusses several power management schemes that aim to minimize the power consumption for sensor nodes.

The rest of this chapter is organized as follows: Section 2 discusses the energy management issue in WSNs, whereas Section 3 presents the energy management strategies that can be adopted to minimize the power consumption for sensor nodes in the WSN. And, finally, Section 4 concludes the work presented in this chapter.

2. Energy management in WSNs

This section discusses the main energy management considerations when designing or developing an algorithm for WSNs. In general, a sensor network consists of a sensor nodes linked to each other using wireless communication protocol. Usually, a sensor network involves various types of nodes with different capabilities (memory size, on-board battery capacity, and processor speed). For instance, **Figure 2** shows a sensor network with three different types of nodes (coordinator, router, and end-device) that exist in the ZigBee communication protocol. Usually, a single coordinator is required to start and coordinate the WSN, whereas a number of active routers are required to forward sensed data in the WSN through multihop communication, and a large number of end-device nodes are expected in the WSN, where end-device node may go to sleep mode.

According to the different energy consumption levels in the WSN based on the type of sensor node that employed in the area of interest, it is important to study

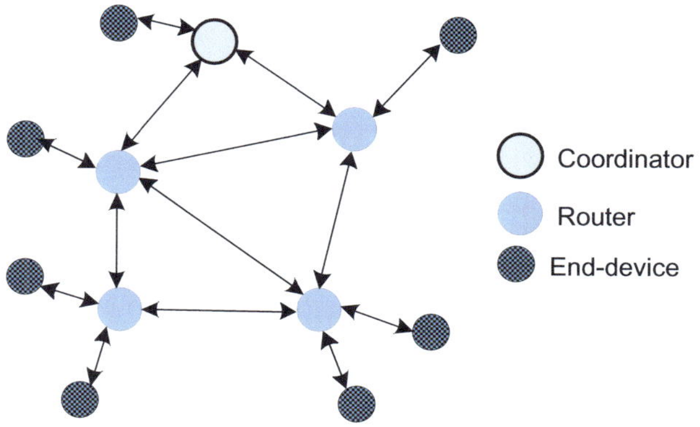

Figure 2.
Mesh WSN with three different sensor nodes.

the field of energy management to minimize the power consumption for sensor network. As presented in **Figure 3**, the energy management is based on two main considerations: energy consumption and energy provision. The former focuses on the operations and devices that deplete the energy through performing transmission, reception, and data processing, whereas the later intends to discover different methods for supplying the sensor node with the required energy source in order to allow the WSN to operate as long as possible.

The energy provision is further classified into: battery-driven, transference, and harvesting. The battery-driven classification is based on the deployment of a battery source for powering the sensor node, whereas the battery might be replaceable, fixed, or rechargeable. The transference classification employs such methods for transferring energy from the source to the sensor node (destination), for instance, the employment of microwaves and radio frequency energy. The harvesting-based classification uses for instance energy from solar, wind, thermal, etc.

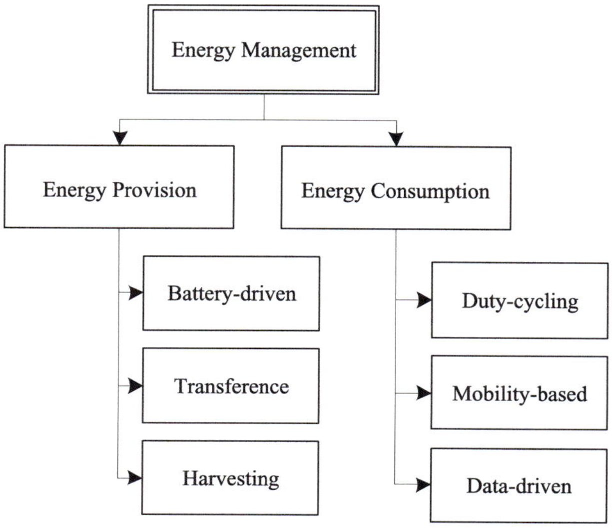

Figure 3.
Energy management in WSN.

On the other hand, there are huge efforts have been made to design and implement an efficient energy management schemes to save the limited energy available for each sensor node. The energy consumption can be further classified into: duty-cycling, mobility-based, and data-driven. Through the duty cycle method, the sensor node can alternate between sleep and active modes in order to minimize the power consumed in the active mode. In the mobility-based method, a mobile node is employed to collect the sensed data from stationary sensor nodes, and therefore minimize the power consumed in multi-hop forwarding of data. The data-driven methods are based on prediction and aggregation algorithms to minimize the power consumed in the transmission process.

3. Energy management schemes in WSNs

Energy management involves saving the onboard energy of sensor nodes in order to allow the sensor node to operate for the maximum lifetime possible. Through studying and analyzing the available literature, it is important to categorize the energy management schemes into four main categories as presented in **Figure 4**.

3.1 Battery management schemes

Battery management includes exploiting the internal characteristics of batteries to evoke their charge in order to maximize the sensor node lifetime. Therefore, in this section the battery management schemes are considered in two ways of views: node energy management and energy balancing.

Node energy management aims to allow sensor nodes to operate permanently in the WSN. Authors of [11] explored the Dynamic Power Management (DPM) strategy in WSNs that established the sleep and active modes for power management. DPM minimizes the energy consumption for each sensor node with the help of micro-operating embedded system. Moreover, several research works [12–17] have focused on the DPM in order to reduce the energy consumption for each sensor node, hence maximizing the WSN throughput.

The energy balancing schemes on the other hand achieve a balance between the energy generation and the energy consumption. For longer WSN lifetime, the efficient and balanced power consumption is highly important. For instance, authors of [18] presented a solution for insufficient energy problem in the sensing unit and excessive usage of power in transmission unit for sensor nodes through setting up a decent harmony among them to prolong throughput up to some extent. In addition, several research works have focused on the energy balancing scheme [19–22].

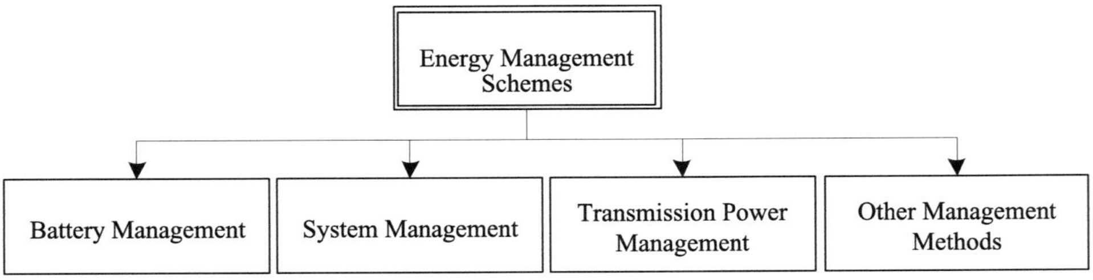

Figure 4.
Classification of different management energy methods for WSNs.

3.2 Transmission power management schemes

In general, data transmission is considered as the most power consumption module with comparison to sensor node's other modules. The transmission power management schemes can be categorized into three main categories as follows: Medium Access Control (MAC) layer management, routing policies.

The MAC layer management schemes are adopted the MAC protocol to minimize the power consumption for the sensor nodes. MAC protocol is considered as the bottom segment protocol for network communication in WSNs. Several research works [23–27] have explored the divergent MAC protocols for enormous applications of WSNs.

The routing protocols on the other hand aim to set up the best link between the source node and the destination node, without compromising some major performance character tics. Routing protocols focus on the power saving, where several routing protocols and systems [28–39] have been developed and implemented for forwarding data in WSN to reach the destination or the sink node.

3.3 System management schemes

The system power management schemes are accomplished in the processor unit using power and device management strategies. The substantial dropping in power consumption offers efficient hardware design. Moreover, the power consumption might be further minimized by some other features including turning-off the sensor node over idle situations or operating in power-saving mode. The system management schemes involve processor power management and device management.

Generally, the power consumption of the sensor node's processor is affected by several parameters including: processor clock speed and the number of command executed per unit time. The processor power management strategy tries to minimize the number of performed calculations and the processor's power consumption. Several research works [40–42] have adopted various power management methods, for instance: employing the power saving mode to minimize the power consumption of a certain sensor node in the WSN.

On the other hand, using intelligent mobile sensor nodes, the power management can minimize power usage considerably. The design and development of the sensor node hardware have been proposed for device management schemes, which minimize the energy consumption. Through this management technology, the intelligent device employs an operating system that aims to reduce the power consumption using various power saving modes according to the sensor node's energy usage. Several device management systems for WSNs have been proposed recently [43–48] with various functionalities and outcomes.

3.4 Other power management schemes

This subsection discusses other WSN management systems including: load balancing, duty cycling, and mobility-based systems.

Load balancing includes managing power usage of the transmission segment. Several data clustering approaches [49–53] have been developed to extend the WSN lifetime and enhance the network throughput, where a cluster head is elected in order to collect, aggregate, and then transmit the sensed data to the base station. In general, cluster-based approaches significantly minimize the power consumption for WSNs.

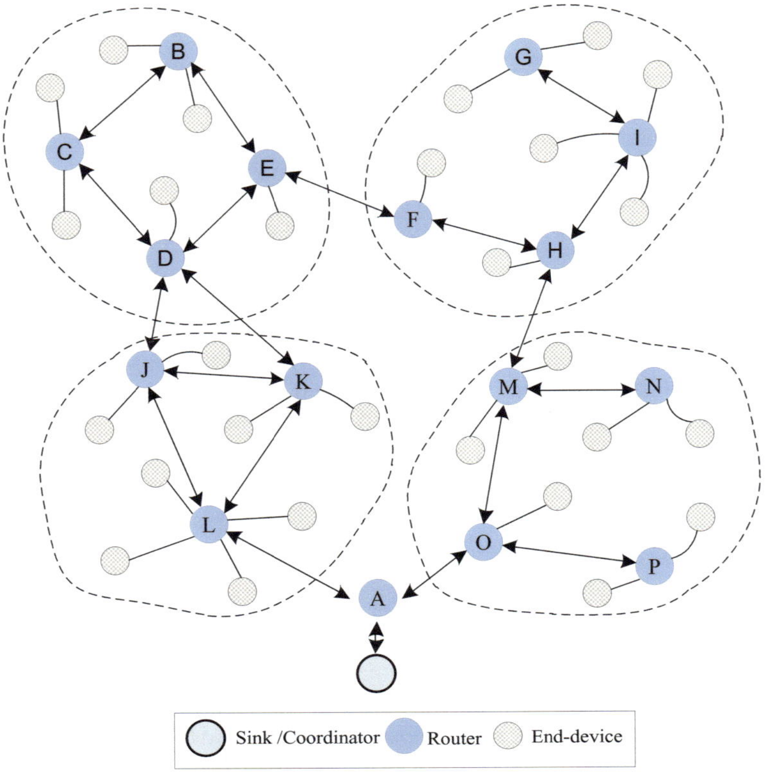

Figure 5.
Clustering concept in WSN.

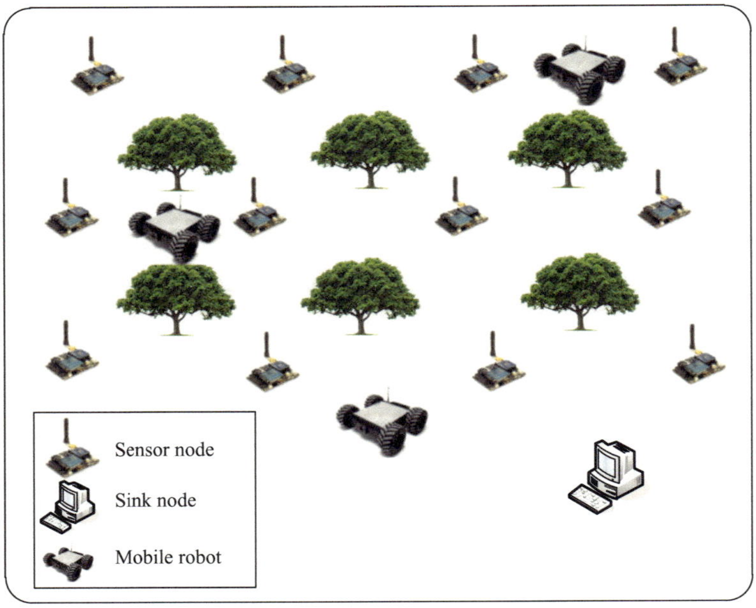

Figure 6.
Employment of mobile nodes to collect the sensed data.

Figure 5 shows the concept of dividing the sensor nodes in a WSN into groups, where a cluster head is selected to each sensor group. The role of the cluster head is to collect data from sensor nodes in its group, aggregate, and transmit the collected data to the sink node (base station).

Duty cycling management schemes manage the power consumption to extend the WSN lifetime. Duty cycling approaches play a key role in enhancing the energy consumption and the WSN lifetime. Several algorithms have been proposed [54–59] that estimate the duty cycle for each sensor node by switching among wakeup and sleep modes in order to minimize the total power consumption for each sensor node.

The mobility-based approaches consider employing mobile sensor nodes to attain energy conservation in the WSN. In WSNs, the mobile nodes are employed to collect the sensed data from stationary sensor nodes distributed over the area of interest. Many research works [60–68] have conducted employing mobile sensor nodes in their studies, with the aim of minimizing the power consumption for fixed sensor nodes and minimize the multihop commination over the WSN. **Figure 6** presents the concept of employing a mobile robot node in the WSN field.

4. Conclusion

A WSN is made up of a set of sensor nodes that are supplied by batteries to operate and interconnected using radio links to guarantee reception, processing, and transmission. Energy consumption is a critical issue in WSNs. Various significant challenges have been overcome to maximize the WSN lifetime, and hence increase the WSN throughput. This chapter discusses several energy management solutions for WSNs, ranging from deployment and connectivity to routing and securing information. In this chapter, the energy management schemes were divided into four main categories: battery management, system management, transmission power management, and other management schemes. Each energy management scheme was discussed, in addition to presenting several research work that support the discussed energy management scheme.

Author details

Tareq Alhmiedat
Industrial Innovation and Robotics Center, University of Tabuk, Tabuk, Saudi Arabia

*Address all correspondence to: t.alhmiedat@ut.edu.sa

References

[1] Khan JA, Qureshi HK, Iqbal A. Energy management in wireless sensor networks: A survey. Computers & Electrical Engineering. 2015;**41**:159-176

[2] Mansour S, Nasser N, Karim L, Ali A. Wireless sensor network-based air quality monitoring system. In: 2014 International Conference on Computing, Networking and Communications (ICNC). New York: IEEE; 2014. pp. 545-550

[3] Alhmiedat TA, Omar F, Taleb AA, Alsswey A. Road safety and energy saving proposed system: A Zigbee WSN approach. International Journal of Online Engineering. 2015;**11**(2):55-59

[4] Alhmiedat T, Ghassan G. A low cost zigbee sensor network architecture for indoor air quality monitoring. International Journal of Computer Science & Information Security. 2017;**15**(1):140-144

[5] Alhmiedat T, Omar F, Taleb AA. A hybrid tracking system for zigbee WSNS. In: 2014 6th International Conference on Computer Science and Information Technology (CSIT). New York: IEEE; 2014. pp. 71-74

[6] Arroyo P, Herrero JL, Suárez JI, Lozano J. Wireless sensor network combined with cloud computing for air quality monitoring. Sensors. 2019;**19**(3):691

[7] Muduli L, Mishra DP, Jana PK. Application of wireless sensor network for environmental monitoring in underground coal mines: A systematic review. Journal of Network and Computer Applications. 2018;**106**:48-67

[8] Alhmiedat T, Salem AA. A hybrid range-free localization algorithm for zigbee wireless sensor networks. International Arab Journal of Information Technology (IAJIT). 2017;**14**:647-653

[9] Alhmiedat T. An adaptive indoor positioning algorithm for ZigBee WSN. In: Fifth International Conference on the Innovative Computing Technology (INTECH 2015). New York: IEEE; 2015. pp. 51-55

[10] Singh J, Kaur R, Singh D. A survey and taxonomy on energy management schemes in wireless sensor networks. Journal of Systems Architecture. 2020;**111**:101782

[11] Zhang Y, Li W. Modeling and energy consumption evaluation of a stochastic wireless sensor network. EURASIP Journal on Wireless Communications and Networking. 2012;**2012**(1):1-11

[12] Jin S, Yue W, Sun Q. Performance analysis of the sleep/wakeup protocol in a wireless sensor network. International Journal of Innovative Computing Information and Control. 2012;**8**(5):3833-3844

[13] Alhmiedat TA, Yang SH. A ZigBee-based mobile tracking system through wireless sensor networks. International Journal of Advanced Mechatronic Systems. 2008;**1**(1):63-70

[14] Kaebeh Yaeghoobi SB, Soni MK, Tyagi SS. Dynamic and real-time sleep schedule protocols for energy efficiency in WSNs. International Journal of Computer Network and Information Security (IJCNIS). 2016;**8**(1):9-17

[15] Shah T, Javaid N, Qureshi TN. Energy efficient sleep awake aware (EESAA) intelligent sensor network routing

protocol. In: 2012 15th international multitopic conference (INMIC). New York: IEEE; 2012. pp. 317-322

[16] Alhmiedat T, Samara G, Salem AO. An indoor fingerprinting localization approach for ZigBee wireless sensor networks. arXiv preprint arXiv:1308.1809. 2013

[17] Alhmiedat T, Salem AO, Taleb AA. An improved decentralized approach for tracking multiple mobile targets through ZigBee WSNs. arXiv preprint arXiv:1307.3295. 2013

[18] Patel S, Sherrill D, Hughes R, Hester T, Huggins N, Lie-Nemeth T, et al. Analysis of the severity of dyskinesia in patients with Parkinson's disease via wearable sensors. In: Proceedings International Workshop on Wearable and Implantable Body Sensor Networks, IEEE Computer Society. New York: IEEE; 2006. pp. 123-126

[19] Babayo AA, Anisi MH, Ali I. A Review on energy management schemes in energy harvesting wireless sensor networks. Renewable and Sustainable Energy Reviews, Elsevier. 2017;**76**:1176-1184

[20] Ni X, Yuan D, Teng Y, Song M. Energy efficient power allocation scheme for multi-cell with hybrid energy sources. In: International Symposium on Personal, Indoor and Mobile Radio Communications—(PIMRC): Mobile and Wireless Networks. New York: IEEE; 2015. pp. 1611-1616

[21] Buwaya J, Rolim J. Bounding distributed energy balancing schemes for WSNs via modular subgames. In: International Conference on Distributed Computing in Sensor Systems (DCOSS), IEEE Computer Society. 2016. pp. 153-160

[22] Salem AA, Alhmiedat T. Energy-efficient clustering WSN system for environment monitoring applications. International Journal. 2020;**8**(5):2126-2132

[23] Ye W, Heidemann J, Estrin D. An energy-efficient MAC protocol for wireless sensor networks. IEEE INFOCOM. 2002;**3**:1567-1576

[24] Jamieson K, Balakrishnan H. Sift: A MAC protocol for event-driven wireless sensor networks. In: European Workshop on Wireless Sensor Networks. Berlin: Springer; 2003. pp. 1-23

[25] Van Dam T, Langendoen K. An adaptive energy-efficient mac protocol for wireless sensor networks. In: Sensor and System. New York: ACM; 2003. pp. 1-10

[26] Sohrabi K, Gao J, Ailawadhi V, Pottie GJ. Protocols for self-organization of a wireless sensor network, allerton conference on communication, computing and control. IEEE Personal Communications. 2000;**16-27**:1002-1006

[27] Kalidindi R, Ray L, Kannan R, Iyengar S, Hall C, Rouge B. Distributed energy aware MAC Layer Protocol For Wireless Sensor Networks Louisiana State University, International Workshop on Wireless Networks 2003. pp. 1-5

[28] Heinzelman WR, Chandrakasan A, Balakrishnan H. Energy-efficient Communication Protocol for Wireless Microsensor Networks. In: Proceedings of the 33rd Hawaii International Conference on System Sciences. New York: IEEE Computer Society; 2000. pp. 1-10

[29] Manjeshwar A, Agrawal DP. TEEN: A routing protocol for enhanced efficiency in wireless sensor networks. In: Proc. of the 15th Parallel and Distributed Processing Symp. San Francisco: IEEE Computer Society; 2001. pp. 2009-2015

[30] Aliouat Z, Harous S. Energy efficient clustering for wireless sensor networks. International Journal of Pervasive Computing and Communications. 2014;**10**:469-480

[31] Bozorgi SM, Shokouhi Rostami A, Hosseinabadi AAR, Balas VE. A new clustering protocol for energy harvesting-wireless sensor networks. Computers and Electrical Engineering. 2017;**64**:233-247. DOI: 10.1016/j.compeleceng.2017.08.022

[32] Kulik J, Heinzelman W. Negotiation-based protocols for disseminating information. In: Wireless Networks. Vol. 8. Kluwer Academic Publishers; 2002. pp. 169-185

[33] Braginsky D, Estrin D. Rumor routing algorthim for sensor networks. In: Proceedings of the 1st ACM International Workshop on Wireless Sensor Networks and Applications—WSNA'02. 2002. pp. 22-23

[34] Intanagonwiwat C, Govindan R, Estrin D, Heidemann J, Silva F. Directed diffusion for wireless sensor networking. IEEE/ACM Transactions on Networking. 2003;**11**:2-16

[35] Karp B, Kung HT. GPSR : Greedy perimeter stateless routing for wireless networks. In: International Conference on Mobile Computing and Networking, Proceedings of the 6th Annual ACM/IEEE. 2000. pp. 234-254

[36] Sanchez JA, Ruiz PM, Liu J, Stojmenovic I. Bandwidth-efficient geographic multicast routing protocol for wireless sensor networks. IEEE Sensors Journal. 2007;7:627-636

[37] Shah RC, Rabaey JM. Energy aware routing for low energy ad hoc sensor networks. In: Proceedings of IEEE Wireless Communication. New York: IEEE; 2002. pp. 350-355

[38] Jakobsen MK, Madsen J, Hansen MR. DEHAR : A distributed energy harvesting aware routing algorithm for Ad-Hoc multi-hop wireless sensor networks. IEEE. 2010:1-9

[39] Alhmiedat T. Low-power environmental monitoring system for ZigBee wireless sensor network. KSII Transactions on Internet and Information Systems (TIIS). 2017;**11**(10):4781-4803

[40] Zheng R, Kravets R. On-demand power management for ad hoc networks. Ad Hoc Networks. 2005;**3**:51-68

[41] Abiodun AS, Anisi MH, Ali I, Akhunzada A, Khan MK. Reducing Power Body Area Networks. IEEE Consumer Electronics Magazine. 2017:38-47

[42] Alhmiedat TA, Yang S. Tracking multiple mobile targets based on ZigBee standard. In: Proceedings of the 35th Annual Conference of the IEEE Industrial Electronics Society. 2009

[43] Shimada H, Ando H, Shimada T. Pipeline stage unification: a low-energy consumption technique for future mobile processors. In: ZSLPED'O3. ACM; 2003. pp. 326-329

[44] Silva A, Liu M, Moghaddam M. Power-management techniques for wireless sensor networks and similar low-power communication devices based on nonrechargeable batteries. Journal of Computer Networks and Communications. 2012;**2012**:1-10

[45] Dargie W. Dynamic power management in wireless sensor networks: state-of-the-art. IEEE Sensors Journal. 2012;**12**:1518-1528

[46] Tsai KL, Ye MY, Tsai SH, Wang YY, Zhuang YH. Attack-resistant power management scheme for wireless sensor network. In: 2015 International Conference on Advanced Robotics and Intelligent Systems, ARIS 2015. 2015. pp. 1-4

[47] Aloulou R, Lucas De Peslouan P-O, Mnif H, Alicalapa F, Luk JDLS, Loulou M. A power management system for energy harvesting and wireless sensor networks application based on a novel charge pump circuit. International Journal of Electronics. 2015;**103**:1-12

[48] Pughat A, Sharma V. Performance analysis of an improved dynamic power management model in wireless sensor node. Digital Communications and Networks. 2017;**3**:19-29

[49] Pan JS, Dao TK. A compact bat algorithm for unequal clustering in wireless sensor networks. Applied Sciences. 2019;**9**(10):1973

[50] Ahmad B, Jian W, Ali ZA, Tanvir S, Khan MSA. Hybrid anomaly detection by using clustering for wireless sensor network. Wireless Personal Communications. 2019;**106**(4):1841-1853

[51] Alhmiedat T. An adaptive energy-efficient data collection system for ZigBee wireless sensor networks. International Journal of Distributed Sensor Networks. 2015;**11**(12):734937

[52] Han Y, Li G, Xu R, Su J, Li J, Wen G. Clustering the wireless sensor networks: a meta-heuristic approach. IEEE Access. 2020;**8**:214551-214564

[53] Bozorgi SM, Bidgoli AM. HEEC: A hybrid unequal energy efficient clustering for wireless sensor networks. Wireless Networks. 2019;**25**(8):4751-4772

[54] Joseph V, Sharma V, Mukherji U. Optimal sleep-wake policies for an energy harvesting sensor node. IEEE Communications Society. 2009:1-6

[55] Zhang Y, Feng CH, Demirkol I, Heinzelman WB. Energy-efficient duty cycle assignment for receiver-based convergecast in wireless sensor networks. In: Global Telecommunications Conference (GLOBECOM) IEEE. 2010. pp. 0-4

[56] Yang F, Augé-Blum I. Delivery ratio-maximized wakeup scheduling for ultra-low duty-cycled WSNs under real-time constraints. Computer Networks. 2011;**55**:497-513

[57] Rout RR, Ghosh SK. Enhancement of lifetime using duty cycle and network coding in wireless sensor networks. IEEE Transactions on Wireless Communications. 2013;**12**:656-667

[58] Shrestha N, Youn JH, Sharma N. A code-based sleep and wakeup scheduling protocol for low duty cycle sensor networks. Journal of Advances in Computer Networks. 2014;**2**:188-192

[59] Cheng L, Niu J, Gu Y, Luo C, He T. Achieving efficient reliable flooding in low- duty-cycle wireless sensor networks. IEEE/ACM Transactions on Networking. 2016;**24**:1-14

[60] Basagni S, Carosi A, Melachrinoudis E, Petrioli C, Wang ZM. Controlled sink mobility for prolonging wireless sensor networks lifetime. Wireless Networks. 2008;**14**:831-858

[61] Kim HS, Abdelzaher TF, Kwon WH. Minimum-energy asynchronous dissemination to mobile sinks in wireless sensor networks. ACM Sensory Systems. 2003:193-204

[62] Khan AH, Jafri MR, Javaid N, Khan ZA, Qasim U, Imran M. DSM: dynamic sink mobility equipped DBR for underwater WSNs. Procedia Computer Science. 2015;**52**:560-567

[63] Taleb AA, Alhmiedat TA, Taleb RA, Hassan OA. Sink mobility model for wireless sensor networks. Arabian Journal for Science and Engineering. 2014;**39**(3):1775-1784

[64] Khan MA, ul Amin N. Energy efficient clustering using fixed sink mobility for wireless sensor networks. International Journal of Advanced Computer Science and Applications (IJACSA). 2016;**7**:505-510

[65] Taleb AA, Alhmiedat T. Depth first based sink mobility model for wireless sensor networks. International Journal of Electrical and Electronics Computer Systems. 2014;**19**:9-14

[66] Vijayalakshmi K, Manickam JML. Mobisink- intelligent mobility pattern based routing protocol for efficient data gathering in large scale wireless sensor networks. International Conference on Control, Instrumentation, Communication and Computational Technologies (ICCICCT), IEEE. 2016:21-25

[67] Taleb AA, Alhmiedat T, Hassan OA, Turab NM. A survey of sink mobility models for wireless sensor networks. Journal of Emerging Trends in Computing and Information Sciences. 2013;**4**(9):679-687

[68] Xing G, Li M, Wang T, Jia W, Huang J. Efficient rendexvous algorithms for mobility-enabled wireless sensor networks. IEEE Transactions on Mobile Computing. 2012;**11**:47-60